CONTENTS

I. GOLD ❈ 1
   The Power of Gold
   The Famous "Gold Standard"
   Gold and the Contemporary World Condition
   Gold as an Investment
   The Future of Gold

II. SILVER ❈ 39
   Properties of Silver
   Uses of Silver
   Silver and American Politics
   The Price of Silver
   Silver Coins
   Investors and the Inventories Question

III. PLATINUM ❈ 51
   The Early History
   Modern Uses of Platinum
   The Esthetics of Platinum
   Platinum as an Investment
   Platinum and Currency Fluctuations

IV. TAX STRATEGIES FOR
   PRECIOUS METALS INVESTORS ❊ 69

CONCLUSION:
   WHY PRECIOUS METALS? ❊ 71

PORTABLE Wealth

The Complete Guide to Precious Metals Investment

Adam Starchild

PALADIN PRESS • BOULDER, COLORADO

Also by Adam Starchild:

Keep What You Own:
  Protect Your Money, Property, and Family
  from Courts, Creditors, and the IRS
Swiss Money Secrets: How You Can Legally
  Hide Your Money in Switzerland
Using Offshhore Havens for Privacy and Profit

Portable Wealth:
The Complete Guide to Precious Metals Investment
by Adam Starchild

Copyright © 1998 by Adam Starchild

ISBN 10: 0-87364-959-1
ISBN 13: 978-0-87364-959-9
Printed in the United States of America

Published by Paladin Press, a division of
Paladin Enterprises, Inc.,
Gunbarrel Tech Center
7077 Winchester Circle
Boulder, Colorado 80301 USA
+1.303.443.7250

Direct inquiries and/or orders to the above address.

PALADIN, PALADIN PRESS, and the "horse head" design
are trademarks belonging to Paladin Enterprises and
registered in United States Patent and Trademark Office.

All rights reserved. Except for use in a review, no
portion of this book may be reproduced in any form
without the express written permission of the publisher.

Neither the author nor the publisher assumes
any responsibility for the use or misuse of
information contained in this book.

Visit our Web site at www.paladin-press.com

FOREWORD: THE RETURN TO PRECIOUS METALS

Investment in precious metals always has been largely a reflection of the investor's perception of the direction of national and world events. While momentary speculators jump in and out of markets hoping for a quick killing, long-term, prudent investors make an effort to understand the fundamental factors that govern economic interaction. They become thoroughly versed in the history, current state, and prospects for their considered investment—whether stocks and bonds, real property, commodities futures, or precious metals.

History supports the premise that at various times investment in tangible commodities is the best protection against the future. It was not mere whim that prompted legendary investors George Soros and Sir James Goldsmith to invest in gold shares and options in 1993. In October 1994, Soros said, "We believe there is now a greater return in the real world than in the in the financial world, and we are moving accordingly. We are taking some of our financial capital and putting it into the real world."

While there is a general economic upswing in the United States and western Europe—and even in jittery Japan, the length and strength of this trend is questionable, and serious factors underscore continued uncertainty: negative real interest rates in the United States and Japan; soaring equity and bond markets;

rising debt; currency instability, including the dollar's continuing weakness against the yen, which has sunk to all-time low levels; the collapse of the much-vaunted European Exchange Rate Mechanism; explosive economic growth in what could be the world's largest national economy—China; and political turmoil and social unrest in Russia and South Africa. Underlying all these is a well-founded nervous apprehension concerning the impending threat of inflation's return. In late 1994, respected surveys of inventory, production, and key economic indicators all gave major cause for concern. As Donald J. Fine, chief market analyst at Chase Securities, Inc., told the *New York Times*, "Under the surface, we see pockets of inflation bubbling."

Only a few years ago some financial "commentators" were telling people precious metals should no longer play any significant part in a balanced modern investment portfolio. Fortunately, millions of wise investors ignored this narrow view, and today investment in gold, silver, and platinum is looked to not only as a hedge against the unknown but as solid insurance against the future, whatever it may hold.

If faltering savings and loans, banks, and brokerage houses are the problem, these are the last places the wise financial decision-maker would want to be left holding paper assets. As government deficits and debt continue to mount, official paper is not much more secure than private paper. With even municipalities defaulting on their bonds, the perceived strength of government paper over private paper has ended. Accordingly, investment banks, mutual funds, and institutional money managers have increasingly followed the Soros-Goldsmith example and invested in the real assets offered by precious metals—gold, silver, and platinum.

As one economist put it, "Man has not yet learned to trust man. You tell me when man is going to change and I'll tell you when the importance of precious metals will cease."

 GOLD

The significance of gold as a highly valuable substance is almost entirely the product of the fertile mind of man. Gold's historic correlation with wealth and power results from man's perception of it, rather than its intrinsic worth. But this in no way denigrates the immense power of the spell this precious metal—as Shakespeare aptly described it, "saint-seducing gold"— has cast over mankind.

Gold as such is only one of many chemical elements—a soft, lustrous yellow, malleable metal assigned by science the atomic number 79; the symbol Au, from the Latin word for gold, *aurum*; part of group 1B of the periodic table, along with copper and silver. Although rare on this earth, gold is found in relatively pure form in nature and is both visually pleasing and easily workable. This purely aesthetic appeal may explain why gold was one of the first metals to attract and hold man's interest.

THE POWER OF GOLD

Most early civilizations—Egyptian, Minoan, Assyrian, and Etruscan—produced elaborate, beautiful golden works of art, and from earliest recorded time gold served as a medium of exchange for goods and services—the first money. As early as

PORTABLE WEALTH

3000 B.C., gold was an accepted symbol of wealth and power among kings and princes over all the known world. In 550 B.C., Croesus, King of Lydia (now western Turkey), ordered the striking of 98-percent pure gold coins; hence the saying, "rich as Croesus," used to describe great wealth even to this day.

The durability of gold is also part of its appeal. When the British archeologist Howard Carter discovered the tomb of Tutankhamen in 1922, the golden treasurers of the Egyptian boy king who reigned from 1361–1352 B.C. were perfectly preserved, including a solid gold coffin 3 millimeters thick that weighed 242 pounds.

And yet, in another sense, there is illogic surrounding gold. Nobody could have ever conceived of a more absurd waste of human resources than to dig gold in distant corners of the earth for the sole purpose of transporting it and burying it immediately in other deep holes specially excavated to receive it and heavily guarded to protect it.

Historically, the predominant human trait associated with gold has been emotion, not logic. Think about it another way: for a person marooned on a barren desert island without food and water, a ton of pure gold stacked on the beach would have a zero value. Gold has no intrinsic worth, other than that which man has given it in the context of his civilization. In other societies, ivory tusks, glass beads, and carved stones have served the same purpose.

The psychology of gold has spawned some of the most gripping works of fiction, the stage, and cinema, chronicling the fabled "gold rush" of 1849 in California and, later, in the Yukon and South Africa. Long before, the Spanish colonial quest for *el dorado* in south and central America drove men to risk everything, including life itself, in their exploration of the unknown New World.

No less a hard-headed realist than a superb student of history, the father of modern France, the late Charles de Gaulle, advocated a return to the international gold standard at a 1956 presidential press conference at the Elysee Palace with a statement that epitomized man's fascination with gold in dramatic terms:

Gold

There can be no other criterion, no other standard than gold. Yes, gold which never changes, which can be shaped into ingots, bars, coins, which has no nationality and which is eternally and universally accepted as the unalterable fiduciary value par excellence.

Gold is indeed the one ultimate method by which nations, whatever their internal politics—socialist, communist, or capitalist—settle their accounts and debts with each other. It is, as M. de Gaulle pointed out, the accepted universal medium of exchange for governments and individuals alike, and the possession or lack of sufficient gold reserves (or their equivalent) may decide—and often has decided—the fate of nations in war and peace.

But with all the admitted utility gold has in human affairs, it is also a substance imbued with deep passion, arousing the best and worst in human character—envy, greed, hatred, anger, sorrow, zeal, love, and joy. When the German archeologist Heinrich Schliemann (1822–1890) discovered a large hoard of golden objects at the sites of ancient Troy and Mycenae, among them a readily identifiable golden death mask, he dramatically told the King of Greece, "I have looked upon the face of Agamemnon." In a more ominous vein, Spain's King Ferdinand gave the following curt order to his conquistadors: "Get gold humanely if you can, but at all hazards get gold."

In time, gold came to be invested with a mystical quality symbolic of divine power. Not only has it adorned countless temples and cathedrals and served as a ritual sacrifice, but ancient legend had it that gold was the child of the god Zeus. When Jason and the Argonauts chased after that golden fleece, they were simply looking for a sheepskin, commonly used in ancient times and as recently as the California gold rush, as a practical filtering device to catch bits of alluvial gold and silver in fast-flowing streams. Then there was man's fascination with the ancient "science" of alchemy, lasting well into the 17th century, seeking in vain the magic formula that would turn dross into gold. This

PORTABLE WEALTH

unfulfilled quest helped fuel the conquistadors and spark the various gold rushes.

A Limited Precious Metal

The greatest early surge in world gold production came after the first voyage of Christopher Columbus. From 1492 to 1600, Central and South America, Mexico, and the Caribbean islands contributed large amounts of gold to world commerce. During the 17th century, Peru, Colombia, Ecuador, Panama, and Hispaniola (now Haiti and the Dominican Republic) contributed 61 percent of the world's newfound gold; by the 18th century they were supplying more than 80 percent.

Following the 1849 California gold rush, North America became the world's major supplier of gold. But by 1890 the gold fields of Alaska and the Klondike in the Yukon edged out the western United States in production, and soon after the southern African Transvaal exceed even these areas.

Today the largest producers of gold are the Republic of South Africa, the United States, Australia, Canada, the Russian Commonwealth of Independent States, the People's Republic of China, Uzbekistan, and Brazil, respectively.

Gold is one of the scarcest and most desired materials on planet earth. Although man has been extracting it for more than 6,000 years, only about 110,000 metric tons actually have been produced, 35,000 tons of which remain in the hands of central banks and other government institutions. If all that gold were brought together in one place, it would be just enough to form a solid 24-carat cube measuring only 18 meters (approximately 55 feet) long on all sides. (The purity of gold is measured in parts of 24ths; thus 12-carat is 12/24 or 50-percent gold; 18-carat is 75-percent gold; 24-carat is pure gold.)

Although gold has a long pedigree, in terms of total production volume it is a modern phenomenon. More gold has been claimed from the earth in the last 150 years than in 6,000 years before. And in recent years the majority of gold production has found its way into private hands—being "hoarded" by those who see it as a bulwark against threats of all kinds.

Gold

Alchemy aside, gold cannot be fabricated by man. Nature limits its supply. All the new gold mined each year totals less than 2,000 metric tons—an amount that could be fitted comfortably into the living room of a small modern home.

Today experts estimate that the world's unmined gold reserves total about 1 billion troy ounces (31 billion metric grams), roughly half of this in the Witwatersrand area of the Republic of South Africa. Major deposits are in Papua New Guinea, Australia, the United States, and Brazil.

To this very day, salvage companies spend millions of dollars hunting hoards of gold hidden beneath the sand in centuries-old, ocean-bottom shipwrecks off the Florida coast and in the Caribbean. And nearby at Cape Kennedy, the American space program depends on multiple technical uses of gold—including the circuitry in the i486 one million transistor Intel computer microchips—in order to launch and sustain its space-bound payloads, both human and satellite.

Because it is noncorrosive and largely nonreactive to other chemicals, gold was one of the first elemental metals (along with silver and copper) used by humans in primitive manufacturing without the need for refining. Minute quantities of gold in plastic or glass sheets serve as excellent insulation from radiation, light, and heat. Gold-plated bearings are used in abrasive atmospheres or where surfaces are exposed to corrosive vapors or fluids. Because it is nontoxic, gold is used in medicine and dentistry where compatibility with human tissue is required.

Gold's malleability allows it to be hammered into extremely thin sheets and drawn into extremely thin wires. An ounce of gold can be beaten out to 300 square feet (about 30 square meters) in thin sheets known as gold leaf.

Gold's other important property is that it is an excellent conductor of electricity and heat. It is this conductive quality that makes gold so important to the electronics industry. Infinitesimal gold circuits on tiny slivers of silicon chips are at the heart of the computer revolution. The Intel i486 microchip mentioned above contains only an infinitesimal amount of gold but sells for much

PORTABLE WEALTH

more because of the flawless performance of the gold circuits, some as thin as 10 microns (0.00004 inch). Gold circuits in microchips are essential in commercial and military aircraft, ship navigation, telecommunications of all kinds, space vehicles, so-called "smart" weapons, and medical diagnostic systems, to name but a few applications.

Bankers also have a deep respect for and attachment to gold—not just coins and bullion, but also the gold in the computer microchips that allow instantaneous financial transfers around the world.

All this gold-related business and industrial activity means constant international demand for gold—and consistently strong gold prices.

Gold as an Economic Shield

Gold is *the* traditional anti-inflation protection.

In almost every national or international crisis—inflation, currency devaluations, stock market plunges, wars, revolutions—gold has been the refuge of choice for affected saints and sinners seeking safe economic haven. When nations clash, citizens of means look for a quick way to protect themselves financially, and gold relieves their apprehension.

In the long run, gold and other precious metals are the most effective protection of personal purchasing power. The Old Testament informs us that during the reign of King Nebuchadnezzar of Babylon (605 to 562 B.C.), one ounce of gold bought 350 loaves of bread. An ounce of gold today will still buy about 350 loaves of bread in Babylon—or anywhere else. Similar historic comparison shopping confirms that the same quantity of gold will buy a loaf of bread in today's London as was required for that purpose in 16th-century England.

This valuable constancy is why so many investors worldwide see gold as the "ultimate asset"—an important part of a total investment portfolio providing security, stability, and, when needed, a loaf of bread.

GOLD

THE FAMOUS "GOLD STANDARD"

In essence the rules for a nation on the classic gold standard required a fixed official price for gold, with gold coin and/or gold-backed freely redeemable paper currency in circulation. (People today seem to have forgotten completely the concept that paper money was supposed to be instantly convertible into gold or silver, rather than just being a government promisssory note.) On the international level, it meant unfettered import and export of gold with all balance of payments deficits settled in gold. Both the internal and external systems, in theory, are self-regulating mechanisms that discipline the economy of the nation and the world. If a nation runs a balance of payments deficit, gold flows out, less is available for internal circulation, prices are controlled or come down, lower prices make exports more competitive, and the balance of payments improves. A nation with a favorable balance of payments receives an influx of gold and the national economy expands. That is the theory, and was in fact the system used by Britain from 1717 to 1919 (with several intervals during which gold convertibility was suspended) and the United States until March 1933, when President Franklin D. Roosevelt ended it.

If you need any proof of gold's preferential status over paper national currencies, consider the half-century-old Bretton Woods monetary agreement adopted by most nations in 1944. Its basic principle was international convertibility of all major currencies into gold or gold-convertible U.S. dollars—a sort of quasi gold standard that used the strongest national currency of that time. The system lasted until August 15, 1971, when, pressed by domestic recession politics and dwindling gold reserves and ignoring economic principle, a totally pragmatic President Richard Nixon suspended U.S. gold payments to foreign governments.

Until August 1971, the U.S. Treasury had purchased gold at $35 per troy ounce, a price set in 1934 and always honored until Nixon's decision. This gave the dollar its gold-link value, giving rise to the now quaint phrase "the dollar is a as good as gold." This mechanism was at the heart of the gold/dollar exchange standard, instilling confidence in people and nations holding dollars as their reserves instead of gold, with the knowledge that those dollars were gold-backed and convertible. There had been a "free market" price for gold in London all through the 1960s. In 1968, the first step of closing the "private gold window" was taken by eliminating the treasury dealing in gold at $35 for nonofficial transactions. After that, prices floated freely for nonofficial transactions, rising as high as $43 in 1969. In August 1971, Nixon and other governments raised the official price to $42. Private market prices immediately adjusted to prices above this "base." By 1980, gold had reached an all-time high of $612.38 an ounce, a sad measure of the dollar's true value and decline.

The creeping blight of the American dollar's 1971 position can best be illustrated by stark numbers. During the 1960s and 1970s, the U.S. balance of payments situation rapidly deteriorated until the value of dollars held by foreigners exceeded U.S. gold reserves $25 billion to $11 billion. This meant that only 44 percent of the overseas circulating U.S. currency was actually backed by gold, hardly an inspiration to those stuck with weakening dollars.

Measured against the period since 1971, Bretton Woods was indeed a comparative "golden age." Consumer prices more than doubled in America between 1944 and 1971, an annual average rise of 3.2 percent, but after the Korean War (1950–1953), the average rise was only 2.3 percent. Of course, there were government price controls imposed at various times during the earlier period. By contrast, since the 1971 Nixon decision, prices have multiplied three and one half times, an annual average rise of 6 percent inflation.

In the broader sweep of history, Bretton Woods is a distant second best to periods of U.S. and world price stability under the classical gold standard, including 1834 to 1862, and 1879 to 1913. American consumer prices varied in a 26-percent range in those

GOLD

62 years, and stood at almost exactly the same level at the start and finish of both periods. From 1834 to 1862, average annual inflation was zero and the annual average variation up or down in prices was 2.2 percent. From 1879 to 1913, when the U.S. and most other major nations were on the gold standard, U.S. consumer prices ranged only 17 percent in 34 years, again with an average annual inflation of zero and annual average price movement in either direction of only 1.3 percent.

This enviable record of gold-based price stability stands in stark contrast to the average price gyrations during and after the American Civil War (6.2 percent from 1862 to 1879), from World War I to Bretton Woods in 1944 (5.6 percent), and after Nixon ended Bretton Woods in 1971 (6 percent).

A strongly related supporting factor for stable international prices in these periods was the Bank of England, privately owned, therefore profit-oriented, and operating on the gold standard. When nations operated a trade deficit with Britain, gold flowed into the bank, which in turn issued increased official British bank notes to be used for acquisition of money-earning assets, boosted imports, and lessened the deficits of exporting countries. Since the pound sterling was freely convertible into gold, the entire world had a true gold anchor in the Bank of England.

So what explains the sharp historic inflationary contrast between periods on the official gold standard and the times on what might better be called the official "funny money" standard?

Essentially, public confidence in the value of money was at the heart of the matter. People were willing to rely upon the standards by which the government issued money—meaning availability of gold coins or gold-backed paper currency that was convertible into gold. Without those standards, public confidence faltered. Gold was the key.

Before 1914 this sort of high-confidence gold-backed money was legally acquired and held by American citizens and banks without any government restrictions. In that year the Federal Reserve System created a new form of money—the

PORTABLE WEALTH

deposits of private member banks held at regional Federal Reserve banks, which served as a substitute for vault cash, against which were issued a new paper currency identified as "Federal Reserve Notes."

At the end of World War I, foreign governments created yet another form of money to which public confidence could attach—"foreign exchange"—mainly gold-convertible U.S. dollar or British sterling assets held by these governments, managed by their central banks, and pledged in lieu of gold and used for exchange to settle international debts. This mutual exchange of paper promises was then treated as a reserve on the books of the central bank, supposedly just as valuable as a gold reserve.

In October 1933, President Roosevelt ordered all privately held gold in the United States confiscated in exchange for paper dollars. Coins having "recognized special value to collectors" and industrial uses were exempted. It was not until 1974 that Congress again legalized the private ownership of gold in the U.S., but the government did maintain a quasi gold standard, pledging to buy gold at $35 an ounce—far below its more recent free market price.

It was a combination gold and foreign exchange standard that was formalized at Bretton Woods. Since 1971, U.S. official monetary reserves have mostly consisted of foreign exchange holdings, a poor substitute for what was once official gold and gold-backed convertible money. Who in his right mind would prefer a dollar backed only by official claims against foreign government assets, as compared to dollars backed by gold in Fort Knox? With the demise of the gold standard, confidence in a firm American dollar has vanished.

People had confidence in a gold-convertible money system because the system worked—by providing real buying power as well as psychological value. The reason for the popular attitude toward gold as a valued object was described by famed economist David Ricardo as the "duality of gold"—a reference to the metal's usefulness both as a governmental monetary unit and as a commercial product in its own right. What makes gold-backed

GOLD

currency preferable (and anti-inflationary) is gold's inherent self-regulating stability.

Under an international gold standard, the total supply of gold coins or bullion (gold in bars, ingots, or wafers of .995 purity) responds to the level of prices in general. In each country, price fluctuations relative to other gold-standard nations produce an in-flow or out-flow of gold or gold-backed money. A worldwide increase in prices and wages discourages gold production because it raises mining and associated production costs, while a fall in prices stimulates more gold output. This means that in the absence of sharp expansions or contractions of credit, or technical changes that radically increase gold mine production, price levels vary only within narrow limits, keeping inflation tolerable at low levels.

Gold convertibility also regulates the total paper money supply and imposes constraints on credit. Free of the gold standard, without this gold convertibility pressure, the American government can and does resort to "printing press" money, expanding credit and inflation at one and the same time. With no present legal requirement to back currency with gold or other precious metals, politicians always seek the quick fix of cheap dollars, looking toward improving their own fate at the next elections.

In 1971, when Nixon chose to take America (and with us, most dollar-dependent nations) off the Bretton Woods quasi gold standard, he moved the United States and the world to a loose "paper dollar standard," based on little more than the daily consensus perception of America's immediate economic prospects. This act marked the beginning of the end of the dollar as the stable pillar of the world monetary system and the start of the currency volatility that has persisted ever since.

Small wonder that the American dollar now trades for less than one tenth of its former official 1971 gold value of $35 an ounce—meaning gold itself commands $350 or more an ounce.

Given the history of economic horrors produced by the continuing U.S. loose money standard, is there any further need to convince a serious investor of the value of owning and holding gold?

PORTABLE WEALTH

You have to choose between trusting to the natural stability of gold and the honesty and intelligence of members of the government. And with due respect to these gentlemen, as long as the capitalist system lasts, vote for gold.

GOLD AND THE CONTEMPORARY WORLD CONDITION

During 1993 the swings in the price of gold were the most extreme in years. From a seven-year low in March 1993 of $326 per ounce, the price rose by August to $407, dropped to $343 in September, and leveled out to $390 by year end—the same price an ounce commanded a year later in October 1994. The 19-percent increase in gold value during 1993 was unique in that it occurred in a largely noninflationary period. But this price is less than half of gold's historic high of $850 in 1980 and still lower than its typical annual peak.

One major reason for these unprecedented noninflation gold price rises has been the basic operation of the economic principle of supply and demand. Largely unnoticed in the Western nations, in 1993 China became the world's largest consumer of gold, a trend mirrored in other Far Eastern countries whose economies are just now preparing to enter the 21st century. A strong argument can be made that today's gold market is being driven by buyers more than sellers, the classic supply-demand market, rather than by fear or inflationary expectations, as some commentators mistakenly believe.

Then too, there were the unsettling difficulties presented by the reunification of the two Germanies, East and West, and the European Community's resistance to the bureaucratic imposition of the so-called "European Monetary System." This short-circuited the European Exchange Rate Mechanism (ERM), causing wary investors to flee from currencies into the safety of gold.

Add to this unusual situation of the recent past the coming inflation predicted by multiple economic indicators, and the possibilities for gold price increases are limited only by imagination.

GOLD

Volatility in gold price historically has been associated with the beginning of a bull market, when investor interest is growing but sentiment has not yet solidified into a trend. Of course, gold has always performed best as an investment during inflationary periods, and there are many international signs that industrialized nations are attempting to reinflate their economies by stimulative monetary and fiscal policies. U.S. money supply (M1 to an economist) showed its largest increase in history from 1990 through 1993, and sooner or later this has to mean increased prices for goods and services. The index of 21 commodities is up 16 percent from its 1993 low, another traditional predictor of coming inflation, and factory capacity is approaching 84 percent. Both the consumer and producer price indices (CPI and PPI) are usually lagging gauges of coming inflation, more a look back than an accurate prediction of the future.

The status of the United States as a debtor nation is always a subject for legitimate concern—and a factor in gold prices. From being the world's largest creditor nation in 1975, the United States had become the world's largest debtor by 1986. But when the debt is viewed as a percentage of a country's true national economic product, nine nations are struggling under current budget deficits greater than that of the United States, including Belgium, Canada, Finland, Greece, Italy, Spain, Sweden, Portugal, and the United Kingdom. The difference is that they don't print U.S. dollars, the largest reserve asset held around the world.

And history has an unfortunate way of not only repeating itself but sometimes stuttering. In 1978 under President Jimmy Carter, the U.S. economy seemed in the best of health, but by 1980 inflation was running around 18 percent, the price of gold was at $825 an ounce, and the index of investment-grade rare coins had appreciated 1,108 percent in what was the biggest bull market of all time. In the face of such a drastic turnaround, the easy way out for any administration could well be more reinflation, a traditional Democratic policy.

Based on the current U.S. government economic policies,

some experts are now predicting the U.S. inflation rate will again begin to rise, with a dramatic slowdown of real economic growth. It is not a coincidence that while the U.S. held more than two thirds (68 percent) of the world's gold reserves in 1950, today it owns only 23.5 percent.

All these factors taken together indicate that the time to latch onto the gold and precious metal anti-inflationary hedge is now. When inflation really hits, smart investors will be taking gold profits while the others are still attempting to figure out what hit them.

A Good Example—A Gold-Backed Currency

Before we leave this discussion of gold as a monetary unit and gold-convertible currency, we should consider the Swiss franc as both an example and a possible gold-related investment you may wish to consider if currency speculation appeals to you.

The Swiss franc is more than a paper currency—it is a currency backed officially by gold. Swiss law requires a minimum 40-percent government gold reserve for the face value of every franc in circulation. Actual Swiss gold reserves amount to a 56-percent reserve, and the franc is valued at the old Central Bank price of US $42.22 per ounce. At today's gold market price, the actual Swiss gold reserves amount to many times the total face value of all Swiss francs in circulation. There is no other national currency in this strong gold-backed position, and this fiscal strength shows in the long-term track record of the Swiss franc during the last half century compared to other paper money, certainly including the dollar.

Switzerland's political and economic stability has contributed to the franc's superior level of performance and its steadily increased value against all other currencies—but so has its gold backing. The Swiss franc has been the world's best investment currency, with interest rates running higher than 10 percent when measured in U.S. dollars.

The enviable Swiss franc record should tell us something about the relationship of currency value, gold backing, and popular confidence in a nation's money system.

GOLD

Good as Gold

For thousands of years, gold has been man's personal primary store of value, more trusted by individuals than any paper investment or printed official currency. Gold cannot be inflated by making more, nor can it be devalued by profligate government decree. And, unlike paper currency or investments such as stocks and bonds, gold is an asset not dependent on anybody's promise of future repayment.

Gold is one of the few investments that has survived—even thrived—during times of economic uncertainty. For example, in the United States following the Wall Street crash and the onset of the Great Depression, from September 1929 to April 1932 the Dow Jones Industrial Index of stocks plunged from 382 to 56—a drop in value of 85 percent—and more than 4,000 U.S. banks were forced to close their doors. Meanwhile, the price of gold actually went up—by around 57 percent.

Gold also increased in value during the events leading up to Black Monday, October 19, 1987, when the Morgan-Stanley index of world stock shares fell 19 percent over 10 days. And during the mini crashes that have afflicted stock markets since then, gold has held its value, ignoring the roller-coaster travails of shareholders' investments.

Historically, gold has been a hedge not only against inflation but deflation as well, serving to reliquefy bankrupt economies at the bottom of recession. Gold may serve in this role again because it has the virtue of being a store of portable wealth not dependent on political promises being fulfilled or the survival of particular nations or domestic political institutions.

Economic history shows that with the singular exception of gold coins, official currency—"the coin of the realm"—inevitably declines in value and purchasing power, regardless of whose face momentarily graces the dollar, franc, mark, ruble, or yen. In an era when most governments casually ignore any responsibility for backing their currency with real assets, it is particularly important for the investor to create his or her own gold "reserve fund"—a small act of faith in lieu of government's abdication.

GOLD AS AN INVESTMENT

Gold ownership is not so much an investment as it is a speculation.

That is because owning gold produces no current interest or dividend income for the owner. When a gold buyer adds the cost of storage, insurance, and a gold installment buying plan, the total is more than the gold's current market value. Speculators who gamble on rising gold prices often lose badly, but that is not because of some inherent defect of gold. Such speculation is far removed from the proper role gold plays in an investment portfolio, as a means of achieving balance, diversification, and inflation insurance.

Small wonder that the best financial and banking experts recommend prudent investors hold a minimum of 5 percent and up to 20 percent of their total portfolio in gold or other precious metals, such as silver and platinum. They do so because they know there is no more reliable insurance than one's personal possession of this "ultimate asset"—always a certain guarantee in the face of the uncertain fiscal future.

To put an entire personal savings program into diversified paper investments without a gold component is to have a truly unbalanced plan. Pure gold should also play a part. Stocks, bonds, insurance, and pension funds are all paper currency-denominated assets that diminish with inflation. Every nation's paper currency buys less than it did a century ago, but gold now buys almost twice as much as it did in 1900. That long-term record proves gold provides true extended inflation insurance, a guarantee far superior to any small profit to be grabbed from overnight conjecture on short-term price trends.

Shortsighted "investors" at times express disappointment with the price performance of gold in recent years, but they forget that gold's primary purpose is to serve as an inflation hedge. Fortunately, American economic inflation has been modest in the past decade, so there has been little immediate need for such protection. But that welcome circumstance is not an indictment of gold as a bad investment, because it is highly misleading to contrast gold with other investment performances in an inflationary sense. Rather, in the spirit of a whimsical Winston Churchill, who dubbed his elaborate plans for his own funeral "Operation Hope Not," owning gold is more closely akin to buying life insurance and not dying.

An investor who pays attention only to the current price of gold misses the point. In the last decade gold played exactly the role it was supposed to play in an investment portfolio—providing a steady store of value with built-in constant inflation protection.

Deciding the Price of Gold

Gold owners may enjoy the glittering presence of their prize, but its value at any given moment is what is more important. While gold will likely always have continuing purchase power, how is the exact price of gold decided on any given day?

Every business day gold is traded around the world and around the clock, with the price fluctuating back and forth between London, Zurich, Hong Kong, Tokyo, Sydney, New York, and other major gold trading centers.

However, daily quotations published in your local newspapers are usually prices issued at noon or at the close of trading by New York's Commodity Exchange Inc. (COMEX), or they may be from the famous twice-daily "fixing" by major bullion dealers at a trading session in the London offices of N.M. Rothschild & Sons.

In a daily ritual faithfully carried out since 1919, each of the five major members of the London Gold Market is represented at the fixings at 10:30 A.M. and 3:00 P.M. (5:30 A.M. and 10:00 A.M. EST in New York), each in direct telecommunication with his own trading room, while a representative of one of the major bullion houses acts as chairman of the assembled group.

After considering the price at which gold has been trading earlier that day, the chairman offers a suggested price that the gold market members communicate to their own traders. The traders respond to the price by telling the chairman whether they wish to buy, sell, or have no interest. The chairman then suggests other prices until all buyers and sellers agree on both price and quantity. At that point, worldwide supply and demand comes into equilibrium and the chairman declares the price "fixed."

But immediately after this daily gold benchmark is established, as a price reference for the world market, local traders are free to vary the price in response to the realities of supply and demand as expressed in their own trading area.

Basic Rules for Investors

There are some basic principles an investor in gold should follow, in addition to the usual prudential judgments to be applied before any investment. These rules will be explained more fully further on in this text, but here is a summary:

The investor should be certain of the good reputation of the gold dealer. Gold bullion should contain marks of a known refiner showing weight and fineness and come with a written guarantee. Ask for a guarantee if one is not offered. Commissions added to the gold price should not exceed 10 percent, and if they do, shop elsewhere. Do not agree to an arrangement that requires you to resell to the seller at a price the seller determines. It is generally preferable not to take personal possession of gold except in small quantities; a bank receipt or other certificate will prove your ownership adequately. If you do want to receive the gold personally, make the purchase in a state with a reduced sales tax. Delaware is particularly popular for buying gold, because there is no sales tax. Avoid buying very small amounts of gold, unless you just want a gift or souvenir, because the price soars for reduced quantities, especially when commissions are added. Check with your tax advisor about any current IRS rules on gold, since they can easily change with little notice.

As an example, a 1992 IRS ruling prompted American coin dealers, who stopped selling bullion products in the 1980s because of burdensome IRS reporting requirements, again to sell bullion. Since a 1982 IRS ruling, it was widely assumed that

GOLD

regardless of total amount, all retail gold bullion sales required a report to the IRS. The cost and burden of that reporting, coupled with increased government audits and rigid but arbitrary enforcement, caused thousands of small coin dealers and metals brokers to abandon the precious metals bullion business, according to the Industry Council for Tangible Assets, which fought for 10 years to change the regulations.

The newer 1992 regulation, Revenue Procedure 92-103, requires the reporting of personal purchases of gold on IRS Form 1099-B only when the total sale equals or exceeds Commodity Futures Trading Commission gold contract sizes. For gold bars, the reportable CFTC contract or sales size is 1 kilo (32.15 oz.) with fineness of at least .995. For example, for 1-ounce gold Canadian Maple Leafs and 1-ounce gold South African Krugerrands, the total reportable sales size is 25 coins. No other gold investment products are subject to IRS reporting. Amounts for silver, platinum, and palladium products are also detailed in this IRS ruling, copies of which should be available at the office of any precious metals dealer.

Methods of Investing

As an enhancement to man's inherent fascination with gold (as though one was needed), there is a certain exhilaration produced by personally participating in one or several of the large number of different methods available for purchasing gold in today's domestic and international gold market.

A purchaser can invest in physical gold by buying gold bullion (bars, ingots, or wafers), gold coins, or medallions—some of which sell for less than $25. Or one might buy a bank-issued gold certificate representing personal ownership that is redeemable in gold upon presentation. Or a prudent buyer can start a gold accumulation program for an original investment of as little as $100, with periodic purchases thereafter on a regular basis.

If you feel more comfortable with stock ownership, you may consider purchasing shares in established gold mining companies in the United States and overseas. There are also gold-based

Portable Wealth

mutual funds, and, for the more speculative investor, gold futures contracts and options.

A word about what you should expect to pay in addition to the current price of gold: in any financial transaction, like exchange of one national currency for another, there is always a difference between the price at which a dealer buys and the price at which he sells. Buyers pay the higher price, sellers receive the lower price. This price difference is often called "the bid-ask spread." In the purchase of stocks the spread can approach 3 percent; in currency exchange, as much as 5 percent. The spread on gold, however, is much smaller and can be as low as one-tenth of 1 percent on very large deals.

Like King Midas of the fabled "golden touch," many investors prefer personal physical possession of their gold, allowing tangible control of their asset. Pride of ownership also moves investors to have their gold within reach so they can display it for friends or just admire it in the comfort of their hopefully security alarm-equipped home. If you find the idea of physical possession of gold appealing, you may want to buy the metal in the form of gold bars or privately minted coinlike commemorative medallions.

Gold Bars

The standard unit of gold in international trading between banks and governments is the 400-troy-ounce (12.5 kilogram) bar of a fineness of 995/1000, with an identifying serial number stamp, traditionally referred to as a London "good delivery bar."

One troy ounce is equal to 1.09714 regular ounce. The troy ounce is a special measurement of gold with ancient Anglo-Saxon origins equaling 31.1035 metric grams of gold. The purest gold of 999.9 fineness is used in the making of these smaller bars.

Gold bullion bars are available in at least 20 sizes and weights, ranging from a tiny 1-gram bar, known as a "wafer," to a pocket-sized kilobar (32.15 troy ounces), to the large 400-troy-ounce London "good delivery bar," valued at $145,000 (at $362.50 an ounce). This variety offers a wide spread in gold bar

GOLD

prices, providing opportunities for investors of small or large amounts of capital.

In addition to the advantages of having gold under one's immediate control, there are several good reasons favoring investment in gold bullion: commissions paid on the buying and selling of bullion are minimal, resale of bullion bearing the marks of reputable refiners is relatively easy, and bullion prices are uniformly quoted throughout the world. Gold bullion also may be purchased from thousands of outlets at precious metals dealers, metals exchange companies, major banks, and many brokerage firms.

Gold Coins: A Universal Treasure

Gold coins have fascinated the world's investors for more than 2,000 years, probably since soon after King Gyges of Lydia ordered the first ones minted in 670 B.C.

There are two main types of gold coins: 1) rare or "numismatic" coins prized by collectors, the value of which is enhanced by their scarcity and is a matter of subjective judgment; 2) bullion or intrinsic coins, always government-issued, which are not scarce and sell at only nominal additional premiums over the actual value of their gold metallic content. It is this second class of gold coins that is the most practical for the investor, since these can be bought and sold easily in varying quantities at a price known worldwide.

Investors purchase gold coins because they provide an easy way to satisfy the need for physical possession of their gold. The popularity of such coins, and privately minted coinlike medallions, can be attributed to their small size, convenient weights, security, and ease of storage and transportation.

In addition to their intrinsic value, many ancient gold coins are both rare antiquities and miniature works of art, bought and sold within the worldwide numismatic collecting community at prices well above the value of their gold content. Rare coins are private and do not require the filing of IRS reports as do gold bullion transactions. Although numismatic gold and gold bul-

PORTABLE WEALTH

lion both increase in value in times of inflation, rare gold coins also increase in value during noninflationary periods of general prosperity. For example, from 1987 through 1989, the price of gold fell from more than $500 an ounce to $350, at a time when a representative index of numismatic gold more than doubled in value.

There are also many collectable government-minted gold legal tender coins issued in limited quantities to commemorate events or persons of national historical importance. The typical gold bullion coin is legal tender within the issuing nation, and its gold content is government-guaranteed. In many cases these coins bear a largely symbolic numerical face value, and their true market value depends totally on the actual percentage of gold content.

Owners of gold bullion coins or privately issued medallions can easily keep track of their current value because most popular gold "pieces" or "rounds," as they are sometimes called, contain one "troy ounce" of pure gold, the price of which is reported daily in the financial section of most newspapers.

Bullion coins of less than a full troy ounce are minted in convenient fractional weights, including one-half, one-quarter, and one-tenth ounces. Among the countries issuing gold bullion coins, now or in the past, are Australia, Austria, the Republic of South Africa, Canada, China, Great Britain, Hungary, Mexico, and the United States.

The purchase price of bullion coins normally includes a 3- to 5-percent premium over the true value of the gold content, but this increase generally will be recovered at resale.

Bullion coins and privately minted gold medallions can be purchased at banks, precious metals dealers, brokerage firms, and jewelers. The smaller gold bullion coins have become an increasingly popular type of investment jewelry, and it is not uncommon to see a person wearing one attached to a matching 24kt gold chain.

Here are some practical tips for a coin collector, especially those new to the trade:

GOLD

- Do not rush into collecting, rather build up your hoard carefully over an extended period of time.
- Be wary of flashy advertising offering alleged huge bargains or discounts.
- Buy uncirculated coins whenever possible; they cost more but are more valuable and easier to resell.
- Watch out for chemically treated coins made to look new or uncirculated.
- Diversify your coin holdings among various nations and avoid fad commemorative coins that may be difficult to resell.
- Get expert advice, especially if you are new at coin collecting.

Here are some of the most popular gold bullion coins available:

American Eagle: A 22-carat (916 fine) gold coin with an American eagle's head issued by the U.S. Mint beginning in 1986 in one-, one-half-, one-quarter-, and one-tenth-ounce sizes, with face values of $50, $25, $10, and $5 respectively. There are also "double eagle" $20 gold pieces struck from 1850 to 1933 containing 0.9675 troy ounce of gold. Because of their scarcity, they often sell for 25 percent or more of their gold value.

Australian Nugget: A 24-carat coin issued privately by GoldCorp of Australia in one-kilogram, 10-, 2-, and 1-ounce sizes, and one-half-, one-quarter-, one-tenth-, and one-twentieth-ounce sizes.

Britannia: A 22-carat (916 fine) legal tender gold coin issued by the Royal Mint in Great Britain struck in sizes of 1-, one-half-, one-quarter-, and one-tenth-ounce sizes.

PORTABLE WEALTH

Sovereign: A British Royal Mint one-pound coin last minted in 1931, and now currently minted as an Elizabeth II sovereign, it contains .2354 ounce of gold. Although it once had a hefty premium, it is no longer expensive.

Maple Leaf: A 24-carat coin issued by the Royal Canadian Mint in sizes of 1-, one-half-, one-quarter-, and one-tenth-ounce sizes.

Peso: This 1947 Mexican 50-peso coin contains 1.2 troy ounces of gold.

Philharmonic: A 1-ounce legal tender 24-carat coin produced by the Austrian Mint, bearing likenesses of various musical instruments, in honor of the world-famous Vienna Philharmonic Orchestra, after which the coin is named. There is also available a restruck 1915 Austrian 100 corona coin containing 0.9802 troy ounces of gold, and a restruck 1908 Hungarian 100 korona with the same gold content, first issued when the two nations were united in the Austro-Hungarian Empire.

Krugerrand: Issued in series since 1967 by the South African Mint, its largest denomination contains exactly 1 troy ounce of gold. It is also issued in one-half-, one-quarter-, and one-tenth-ounce sizes.

One of the best sources for gold coins and bullion is a broker/dealer founded in 1982 by two of the former senior officers of Deak-Perera, at the time the nation's oldest and largest precious metals and foreign exchange firm. The principals of Asset Strategies International, Inc. (Suite 400A, 1700 Rockville Pike, Rockville, MD 20852), Michael Checkan and Glen Kirsch, are not "coin dealers," meaning that they don't take positions in the precious metals, therefore creating a bias to sell certain items. Instead, through their domestic and international network of

GOLD

wholesalers, they buy and sell at competitive prices. They can be called in the United States and in Canada on a toll-free line (800) 831-0007 for current market quotes or general information. Clients and friends of the firm receive its monthly newsletter, *Information Line*, free of charge, to keep them up to date on the precious metals and foreign exchange markets. Asset Strategies International is well known in the financial newsletter industry and, at one time or another, has been recognized as a "recommended vendor" by many writers and publications, including Mark Skousen, Richard Band, Adrian Day, *International Living*, and Taipan. Adrian Day, editor of *Adrian Day's Investment Analyst*, says, "I've frequently recommended Michael and Glen in the past; you can continue to have confidence in utilizing their services." Checkan and Kirsch have been in the precious metals/foreign exchange business for a combined total of 50 years.

Investment without Immediate Possession

If a prospective investor values convenience and ease of ownership transfer over actual physical possession of gold, there are available a wide variety of bank-issued gold certificates, accumulation plans, futures contracts and options, gold mining stocks, and mutual and other gold investment funds.

Issued by many banks, "gold certificates" obligate the bank to deliver a stated quantity and fineness of gold to the holder in accordance with the certificate's exact terms and conditions. Financial institutions issue these certificates in fractional denominations, offering the opportunity to invest in convenient dollar amounts.

Buying gold certificates is a simple process. There usually are no fabrication or delivery charges, and while your bullion is on deposit with the institution, insurance coverage is provided. Regular customer statements provide the approximate current value of your gold investment, and the issuing institution is obligated to deliver that amount of gold at any time you wish, or you can order it to sell the certificate.

PORTABLE WEALTH

Mocatta Gold Delivery Orders

In recent years the purchase of gold bullion and coin "delivery orders" has become a popular form of ownership that avoids the inconvenience and safety concerns that arise with personal possession and storage. These orders are also available for silver and platinum.

A delivery order is a title document evidencing ownership of a specific, serial-number-identified bar of gold or silver bullion, or to specific gold, silver, or platinum coins packaged in a sealed, numbered container unit. At the owner's option, the bullion or coins may be stored either in Zurich, Switzerland, or in Wilmington, Delaware. The depository chosen verifies the deposit by countersigning the delivery order.

Delivery orders are nonnegotiable, transferrable certificates that specifically identify the metals assigned to the order's owner and their physical location. The owner of the order can sell, assign, or collateralize it, but is protected against loss or theft because of nonnegotiability. The metals identified are fully insured by Lloyds of London.

The delivery orders are issued by the Mocatta Metals Group, a worldwide trading consortium founded in 1671 specializing in precious metals with offices in New York, London, and Hong Kong.

The group's London affiliate, Mocatta and Goldsmid, is one of the five firms that officially participate in the twice-daily London gold price fixing, and M&G also chairs the daily silver price fixing.

Delivery orders from Mocatta for gold bullion are available in sizes of one kilobar (32.15 ounces), 100- and 400-ounce bars, and in gold coins including the Canadian Maple Leaf and American Eagle, available in quantities of 10. Silver bullion and platinum bullion and coins are also available.

Buying a Mocatta delivery order means the purchaser owns specific metal identified on the order, which may be inspected at the storage place in Switzerland or Delaware. These metals may be removed by the owner or ordered to be shipped to any destination by endorsing and surrendering the delivery order.

GOLD

Charges for delivery orders include a $100 issuance charge plus one year's storage charge, plus one half of one percent of the total value of the metal purchased, exclusive of broker's sales commission. Usually there is no sales tax imposed.

The Mocatta delivery orders provide an easy method and a high degree of safety, liquidity, and privacy for your precious metals investing. MDOs offer transfer of ownership features that enable the owner to sell, assign, or collateralize the metal easily, yet provide the protection of nonnegotiability, since a lost document can be replaced, unlike a bearer security. Since the order is nonnegotiable, it does not have to be reported if it is taken in or out of the United States. There is no reporting of the purchase to the IRS. An MDO can be issued in the name of a family limited partnership or offshore trust or assigned to one of them as a means of transferring ownership of the metal when the asset protection entity is created.

If you want more information about these orders it may be obtained by contacting Asset Strategies International.

The Mocatta orders are also available in silver and platinum.

Perth Mint Certificate Program

There is a little-known way to hold gold and other precious metals overseas, privately. It's called the Perth Mint Certificate Program (PMCP), and it is an excellent way to ensure your wealth securely, discreetly, flexibly, and inexpensively.

When you buy precious metals in the PMCP, you get a certificate of ownership. The certificate represents a specific item—the bullion or coins you purchased. The Perth Mint sets aside a certain amount of a specific material for use at your sole and personal discretion. The document simply shows "ounce for ounce" what you own that the Perth Mint is holding for you.

The PMCP is an extremely private way to own precious metals. The PMCP is not considered a foreign bank or financial account abroad. Therefore, precious metals stored at the PMCP (in a nonbanking depository) are not reportable, even though the program allows you to transfer wealth from one part of the world to another. The PMCP is not considered a monetary instrument, since it is nonnegotiable and does not provide a payment of a "sum certain" in dollars. In legal terms it is a "warehouse receipt." Your assets and any related documents are stored offshore (in Perth). You retain the ownership certificates, which are transferable but nonnegotiable. In case of an economic catastrophe, you simply use the documents to request delivery from Perth to any number of major financial centers, such as Zurich, London, or Singapore.

Some countries have restrictions on gold ownership, but you may remove your assets from the Perth Mint whenever you wish. When you receive the coins, you can cross country borders without duty (unlike with bullion, which is not always duty-free). There are no import or export duties on precious metals in Australia. Coins purchased in the PMCP enjoy worldwide recognition—you can liquidate them in any major financial market (subject to import restrictions).

The program also allows you your choice of gold, silver, platinum, and palladium. No other certificate program I've researched offers all four metals. And you can sell all or part of your holdings—and receive your proceeds—in a variety of currencies: U.S. dollars, Australian dollars, Swiss francs, or other major foreign currencies. All precious metals transactions are completed based on the London P.M. fix.

The PMCP's products—the highest quality and purity Australian seminumismatic coins—come in various sizes, ranging from one-twentieth of an ounce to 1 kilo (gold, platinum, and silver are available in the 1-kilo size). The PMCP offers low premiums and low storage charges, and there is a $50 certificate charge per transaction. The bigger the transaction, the bigger the savings. Of all the available precious metal buy-and-store programs, the PMCP offers the most inexpensive way to buy precious metals, privately, in a convenient form. Now there is an easy way to hold part of your portfolio in precious metals and get the benefits of global diversification.

For more information on the exclusive Perth Mint Certificate

GOLD

Program, contact Asset Strategies International at the address given earlier. Checkan and Kirsch helped the Perth Mint to design this certificate program using their decades of experience with precious metals and other certificate programs. The Highlights of the Perth Mint Certificate Program are as follows:

- Low certificate fees—US $50 per certificate
- Low storage fees for allocated precious metals—1/2 percent per annum
- *No* storage fees for unallocated precious metals—requires one week's notice for delivery
- Low program minimums—US $25,000 to open, US $5,000 or more to add
- No import or export duties on precious metals in Australia
- Extremely competitive pricing
- Reliability—the Perth Mint is a division of Gold Corporation, wholly owned and operated by the Western Australian government, so your metals are government-guaranteed.
- Discretion—private vaulting relationships overseas require no U.S. government reporting, unlike foreign bank accounts. Storage agreements between the Perth Mint and its clients are coded specifically to the client through the use of a password, client number, and Perth Mint Certificate number.
- Security—Since 1899, the Perth Mint has provided safe storage of precious metals. Storage is primarily available in Perth, which is in Western Australia, one of the most politically and economically stable of the continents. Precious metals are insured (at Perth Mint's expense) by Lloyd's of London.
- Flexibility—there are no predetermined docu-

PORTABLE WEALTH

ment sizes; storage is available for all precious metals (gold, silver, platinum, and palladium) on an allocated or an unallocated basis; delivery is offered in a variety of locations worldwide; PMCP certificates are nonnegotiable, but they are transferable; and storage options can be changed to meet the investor's changing needs.

Gold on the Installment Plan

A serious gold investor also can enjoy all the benefits of gold ownership without the responsibilities and costs of security, handling, and storage simply by using one of the "gold accumulation plans" offered by many reputable precious metals brokerage firms.

Some gold accumulation programs require an initial payment of as little as $100 to start buying gold. Once a personal account is established with a broker, additional investments in amounts as small as $50 or as large as $5,000 are easily accomplished.

Buying gold through accumulation programs offers many advantages, including instant purchases effective that same or the next business day, and since a broker buys and sells in the wholesale bullion dealer market, competitive prices are assured. Investing by the dollar amount rather than by the ounce allows convenient purchases credited in either whole or partial ounces, and discounted commission rates for accumulation plan members are up to 40-percent less than other broker charges for these same transactions.

Absolute safety is another appealing feature. All of the gold is stored in major depositories and fully insured. Record keeping is automatic, and confirmation of each transaction is rendered in a periodic summary statement. Gold allowed to remain in an accumulation program is not liable for taxes until sold at a capital gain, or in some European nations liable for a value added tax (VAT). Both can be avoided by buying gold in offshore markets with no such taxes. An owner can liquidate his or her gold plan holdings at any time and at sale will avoid costly assaying fees for weight and purity testing.

GOLD

Gold Futures and Options

A futures contract is a legally binding contract to buy or sell a designated quantity of a described commodity at a specific time in the future, at a price agreed upon at the time the contract is agreed to. Obviously, such an arrangement involves a gamble for both parties, buyer and seller.

Gold "futures contracts" were originally designed to protect large industrial users (also known in this context as "hedgers") from adverse fluctuations in the price of gold by guaranteeing them a steady gold supply at an established price. Hedgers are companies involved in the production, processing, or marketing of gold—miners, smelters, fabricators—who employ gold future contracts as a practical tool for financial and planning management. As with futures contracts for any commodity, a gold futures contract is a promise on the part of the purchaser to buy or sell a specified quantity and grade of gold on a future date for a certain price.

Most futures contract buyers can be called "speculators," members of the public interested mainly in price fluctuations and possible profits, buying futures when they conclude gold prices will ascend, and selling futures when price declines look likely.

Through the expert use of "leverage"—the difference in amount between the "margin," meaning the amount of capital a speculator is required to deposit in the futures account (10 to 20 percent) and the value of the assets controlled by contract on the operative date—profit potential can be very high, as is the risk.

Several American commodities exchanges, like COMEX in New York and the Chicago Board of Trade, deal in gold futures, usually in contracts for 100 ounces, with established customers required to deposit margins of about 10 percent and new customers as much as 20 percent until they become better known. Some spot contracts are for the London "good delivery" bar amount of 400 ounces.

Compare the futures contract with a simple option, which is not a promise to buy or sell, but rather gives the option holder the right to buy or sell gold at a fixed price on some future date if he or she wishes to do so.

The risks are even higher for those unschooled in the intricacies of options and futures contracts. Those interested in possible investment in gold futures contracts and options should discuss their ideas with a trusted, experienced broker before becoming involved in these most speculative methods of participation in the gold market.

Investing in Gold Mining Stocks

Compared with direct investment in gold itself, buying shares of gold mining company stock will be advantageous or not, depending on your investment needs. Among the advantages is the fact that gold company stocks almost always pay a dividend, often yielding good current income for your investment. Gold metal price increases mean your earnings and dividends will also rise.

It makes a difference whether the shares purchased are in a company completely in gold activity, or whether there is some degree of corporate diversification. The latter situation may diminish the movement of share earnings correlated with gold price changes but may also provide continued cash flow and alternative earnings when gold prices are lower. Individual portfolio requirements will dictate which factors are most important.

Any investment in gold mining stocks should be monitored on a regular basis because known ore sources can be depleted, unit production costs can equal or exceed market prices, and labor problems can shut down the best operating mines.

With gold, as with any investment, there is an ongoing need for timely portfolio supervision. If you would rather not do this yourself but still want to get into gold, a gold-oriented mutual fund with mining and production company shares can do the job for you. Joining a gold investment mutual fund gives the investor a professionally managed safer diversification of gold mining stocks without the need for direct personal investment.

In addition to buying into a U.S.-based gold mutual fund, it ns also possible to invest in an offshore gold mutual fund. Traditionally, offshore mutual funds have had to be bought directly from each fund. But an international discount brokerage organization has now set up a program in which more than 80 offshore funds are available through your discount brokerage account (and, of course, they can also buy gold mining stocks for your offshore account). They offer a "mini-trust" package, which includes a discount brokerage account linked to an offshore bank credit card for instant cash access. For more information, please write to The Harris Organization, Attn: Traditional Client Services, Estafeta El Dorado, Apartado Postal 6-1097, Panama 6, Panama.

The firm is able to correspond in English and Spanish, has a staff of more than 150 employees, and has an equity capital of more than $25 million, which is larger than many of the tax haven banks and trust companies.

Gold Mutual Funds

For those wishing to invest modestly in gold mining stocks with the greatest safety and ease, the answer lies in mutual funds. A phone call to a mutual fund will get you a prospectus, and if you like what you read, you can send in your application form and check, sometimes for as little as $250! Joining a gold mutual fund reduces your personal peril and instead spreads the risk through ownership of shares in many different gold mining companies. And there is also the good possibility of dividends.

Just remember that all investments in gold mining, production, and fabrication are only investments in a business—not in gold itself—and thus carry the normal risks of any stock investment, including the possibility of poor management. Don't let gold blind you. Always check the corporate track record before you leap.

A Special Plan—SwissGold

"SwissGold" is the name describing a modern method of investing in gold with ease and assurance. It is a system based squarely on the protective, anti-inflation insurance aspects of gold, with none of the risks associated with gold contract futures or coin collecting gimmicks. And SwissGold is more efficient and economical than buying gold coins through dealers for their bullion content value.

SwissGold is a special investment account created by the respected UeberseeBank of Zurich, a medium-sized Swiss bank specializing in sound investment management. The bank does not engage in general commercial banking or lending to corporations or foreign governments, so it has no exposure to the risks inherent in such loans—and no conflicts of interest in managing investors' money for maximum results.

Founded in 1965, the bank now serves more than 12,000 clients, managing funds of almost US $3 billion. It is a wholly owned subsidiary of American International Group Inc. (AIG), one of the largest insurance holding companies. AIG has assets exceeding US $45 billion and capital of US $8.3 billion. It employees 33,000 people in more than 130 countries.

SwissGold is based on cost averaging, rather than on trying to outguess the market. It is designed for simple and systematic savings—for example, an investor might decide to put $250 per month into gold. That $250 is going into gold every month, regardless of what the market does. In the long run, the gold acquisition cost will be less than the average market price in the same period. This is called cost-averaging. It requires no market expertise on the part of the investor—just the dedication to make the same fixed investment each month, regardless of the market. (In fact, some investors make a point of *not* looking at the market price.)

A similar technique is used by stock market investors; the cost-averaging principle is the same regardless of what is being bought. They invest a fixed dollar amount every month, rather than buying a fixed unit such as one share or one ounce.

UeberseeBank handles the SwissGold accounts, sending detailed statements on each purchase of gold made for the investor. By purchasing in this manner, investors benefit from

GOLD

the bank's ability to buy at wholesale prices normally available only to large purchasers. In turn, the investor pays no extra fee on small unit amounts nor the regular spread charged when buying and selling gold. The savings can be as much as 3 percent because of the wholesale price and another 8 percent by avoiding small order surcharges. When added to the 20-percent savings that is often typical with cost-averaging, the investor is able to build the gold portion of his portfolio in the most economical way.

Naturally, such accounts are treated with the same secrecy as any other Swiss bank account. Each investor's gold is held separately by the bank acting as trustee in a fiduciary relationship. This arrangement is legally significant because it means the amount of gold in the SwissGold account is always the investor's property, not merely a gold-denominated paper obligation of the bank. Thus solvency or credit standing of the bank cannot affect the investor's holdings, although a bank failure in Switzerland is almost unimaginable even with a commercial bank—and UeberseeBank does not even assume commercial risks.

Of course, the gold is insured as well as securely guarded, and the investor can choose to have it stored in Switzerland, the United States, or Canada.

SwissGold accounts can be tailored to an investor's needs, and flexibility is the key word. An investor can suspend gold purchases at any time without penalty, and account possibilities range from monthly purchases to large lump sum purchases, depending upon the individual investor's needs.

Should you have any concern about investing in gold through a Swiss bank like Uebersee, you should know that in Switzerland banking is a national source not only of income but of great pride. Operating in a country less than half the size of the state of Maine, Swiss banks control more than $400 billion in assets, making the country the third-largest financial center in the world after London and New York.

For people with money to protect, whether a little or a lot, Switzerland is traditionally considered the world's safest repository. These days, the Swiss can give Americans many reasons to leave funds in Switzerland, and SwissGold is one of them, but the promise of total secrecy in financial matters remains one of the greatest attractions of Swiss banks.

Swiss statutory civil law protects a customer and the customer's financial dealings as part of the individual's legal right to privacy. As Article 28 of the Swiss Civil Code, this law not only protects such information but makes the person violating secrecy liable to pay damages to the customer. In addition, the banking law makes it a criminal offense for a bank or any of its employees to divulge information about a customer, punishable by fine or imprisonment.

Information on SwissGold accounts may be obtained from JML Jurg M. Lattmann AG, Swiss Investment Counsellors, Baarerstrasse 53, Dept. 212, CH-6304 Zug, Switzerland.

The firm specializes in assisting American and English investors, and everybody speaks your language.

THE FUTURE OF GOLD

Why should anyone invest in gold at this point in history? Does it, or any other precious metal, really have a useful role to play in the immediate world economic future—and what about the potential for personal profit, much less the need for the traditional gold protection-against-inflation factor?

The answer is fairly simple: gold, and other precious metals, could become the de facto world currency again rapidly, as all other paper money falters and dies in value. There is no real necessity for a formal reinstitution of the classic international gold monetary standard, or even a Bretton Woods-style mixed gold-exchange standard. In the face of financial turbulence and international economic stagnation, gold may very well assume automatically the role of an unofficial monetary safe haven.

However weak the U.S. dollar, unhinged from its gold backing, has been, that dollar still has set the pace in international trade, economics, and banking for the last several decades. But

GOLD

now the U.S. national debt exceeds $4 trillion, headed for $5 trillion, and 61 percent of all federal personal income taxes goes just to pay interest on that debt. Accumulated American international debt now surpasses an unprecedented $500 billion and is rising steadily, and many experts flatly state there is no way the U.S. economy can generate the kind of economic surplus necessary to service this debt. Meanwhile, domestic government attempts to reconcile the contradictory goals of short-term economic stimulation and the long-term reduction of the enormous federal debt. Sooner or later, America, and the world, must face the fact that the United States is insolvent, with all the attendant ramifications of that realization.

Based on history, we know that when inflation occurs investors fear the destruction of the purchasing power of their money; in deflation, they fear default on the part of the issuers of paper assets. In either situation, gold has no peers as an economic refuge.

With all the problems that lie ahead, it is almost inevitable that in the near future, certainly before the turn of the century, there will be a flight to quality money and liquidity—resulting in the re-emergence of gold as the only strong quasi-currency acceptable to all peoples and nations.

Lastly, a word about the price of gold.

Economic activity tends to move cyclically over a period of years. From 1973 to 1982, the return on stocks and bonds averaged only about 6 percent a year, while annual inflation averaged nearly 9 percent. Real net gains were therefore impossible to achieve, but due to a lack of alternative investment possibilities, tangible assets like gold and other precious metals zoomed upward in value.

From 1983 through 1993 the situation was reversed. The nominal return on stocks and bonds set historic levels while governmental restraints reduced inflation to an annual average of 3.8 percent. This meant solid real returns, while the need for and value of tangible assets like gold declined.

Gold, now trading in the neighborhood of $400 an ounce, is

already up over 20 percent from a year ago, but is still at only half of its historic 1980 high of $800 an ounce.

The relatively low price of such tangible assets may not last very much longer. Now, not later, is the time to invest in gold and other precious metals. Tomorrow may be too late.

SILVER

Throughout much of recorded history, the leading use of silver was as official bullion reserves and coinage. At least by the 8th century B.C., all the nations of the ancient Middle East between the Indus and the Nile rivers were using both silver and gold as coins, a governmental monetary standard that eventually came to be called "bimetallism." By 3500 B.C., the Egyptians were using silver without the need for complex refining methods because of its purity. Many silver ornaments have been found in royal tombs of the period.

Even though gold coins were used in this early time, silver and copper coins were far more prevalent among ordinary individuals. Gold might buy a loaf of bread, but because of its greater value, only a very small amount was needed for such a transaction. Silver was much more widely used as a common standard of value. Until the 19th century, most nations were on an official silver or bimetallic standard, with the two metals valued in a ratio of 15:1 or 16:1.

It was not until the discovery of gold, first in Brazil in the 18th century, then in Russia, California, Australia, and South Africa in the 19th century, that gold came to replace silver as the sole monetary standard in most nations.

Platinum group metals generally are called "noble metals"

due to acid resistance. Legally, gold and silver are considered precious. In *some* jurisdictions, platinum, palladium, ruthenium, osmium, and iridium are considered precious as well.

PROPERTIES OF SILVER

Silver is widely found in nature, but the total available amount is quite small compared to other metals. Its scarcity is what classifies it as one of the few "precious metals."

Silver is a heavy metallic element with a brilliant white luster; hence, its chemical symbol is **Ag,** derived from the Latin *argentum*, meaning "white and shining." Silver has the highest known electrical and heat conductivity of any metal. It is used in the manufacture of printed electrical circuits and as an electronic conductor coating, often alloyed with gold and copper.

Unlike gold, silver is usually not found in a pure form but compounded in ore mixed with lead, copper, or zinc. Its availability thus often depends on mining of other metals, with silver as a refining byproduct. Techniques of smelting appearing during the Bronze Age about 3500 B.C. made possible the extraction of silver as a by-product of the refining of these other metals.

USES OF SILVER

Silver has long been used in the manufacture of jewelry, solid silver, and silver-plated objects such as eating utensils (silverware) and other tokens of wealth. Silver is mixed with copper alloy for greater toughness in jewelry and coinage. Its proportion in these alloys is stated in terms of fineness, meaning parts of silver per thousand of the alloy. Sterling silver contains 92.5 percent silver and 7.5 percent of other metals, usually copper, expressed as a fineness of 925. Jewelry silver is usually 80 percent silver and 20 percent copper, or 800 fineness. Silver is often mixed with gold (known as "yellow gold") for jewelry and dental alloys.

Silver also produces a number of compounds, or salts, which are used extensively in photography, including x-rays; as light-

SILVER

sensitive materials in photographic papers, films, and plates; in silvering mirrors; and also for cauterizing wounds and in medical treatments for the eyes and diseases of the skin. About 30 percent of annual current production goes for photographic uses, making this a source of consistent demand. By the 1960s, the increased modern demand for silver for these industrial uses far outstripped the use for coinage and exceeded annual world production, driving its price up.

Today, only about 6 percent of total production goes for coinage. Around 37 percent of silver use is in jewelry, silver plate, and sterling ware, which collectively account for the largest end use.

Silver's resistance to oxidization makes it valuable in critical electrical contacts, switches, solders, and many forms of electrical and electronic equipment. Silver is used in batteries for cameras, calculators, watches, and other small electrical devices. It is also used for bearings in airplanes, autos, diesel engines, and the aerospace industry.

Currently, major uses of silver are for jewelry and silverware (37 percent), photography (30 percent), electronics (13 percent), and coinage (6 percent).

The most important original silver mines were discovered by the Spanish explorers in Central and South America in the 16th century. Later major quantities were found at the Comstock Lode near Virginia City, Nevada, in 1859, and in northern Idaho. In the first 20 years of mining, the Comstock Lode's high-grade ore produced $306 million worth of both silver and gold, $130 million of it gold.

The largest active North American mines are now located in the state of Nevada (16 percent of 1993 world production), in Hidalgo, Mexico (19 percent), and in the Canadian provinces of Ontario and British Columbia (10 percent). Silver is also produced in Peru (13 percent), Russia, Australia, China, Bolivia, Chile, Papua New Guinea, Indonesia, and 70 other countries, which together produced 41 percent of 1993 silver production.

PORTABLE WEALTH

SILVER AND AMERICAN POLITICS

The United States officially adopted a bimetallic money standard in 1792, using both gold and silver as coinage. The dollar was defined as 24.75 grains of fine gold or 371.25 grains of fine silver, a 15:1 ratio that made an ounce of gold worth $19.42. But this official ratio overvalued silver at the mint, since the ratio was set at 16:1 in England, and silver proceeded to drive gold out of circulation until 1834, when gold was revalued at 23.2 grains and the price rose to $20.67. That overvalued gold and drove silver out of circulation; the old saying is true—bad money drives out good money. As a result of the Civil War and its aftermath, from 1862 to 1879, paper U.S. currency unredeemable in either metal was known derisively as "greenbacks," a reference to the bill's reverse-side color—its only "backing."

With the discovery of immense gold deposits in California in 1849, gold prices slipped, while silver increased, with silver eventually being sold on the free market at prices higher than the official values set at the U.S. Mint. This resulted in the adoption by Congress in 1873 of a law (called by many "the Crime of '73") forbidding further silver coinage. In 1878 the Bland-Allison Act ordered the secretary of the treasury to buy silver as an official reserve for currency backing, a reflection of growing populist sentiment.

Later 19th century discoveries of silver in Nevada, Idaho, and other western states again changed the supply and price picture. In reaction, the "silver lobby" finally pushed through the Sherman Silver Purchase Act of 1890, requiring the federal government to buy 4.5 million ounces of silver each month, with official U.S. Treasury notes again to be redeemable in either gold or silver at a ratio of 16:1.

The result was a passionate national economic debate that seriously split the country between Eastern seaboard business interests, who were "tight money" gold supporters, and so-called "free silver" advocates of the West and Midwest—mortgaged farmers and other debtors and silver mine owners—who advocat-

SILVER

ed a return to official bimetallism that would increase (meaning inflate) the money supply in circulation.

The leading and most famous advocate of this "populist" free silver position was William Jennings Bryan of Nebraska, who captured the Democratic Party presidential nomination in 1896 after an electrifying convention speech in which he intoned this memorable defiance of Eastern monied interests:

> *We will answer their demand for a gold standard by saying to them, you shall not press down upon the brow of labor this crown of thorns—you shall not crucify mankind upon a cross of gold.*

Bryan and "free silver" lost to the pro-gold business choice, William McKinley of Ohio, in a relatively close popular vote of 6.5 million to 7 million, a reflection of the bitterly contested monetary issue. In 1900, Congress officially affirmed a single gold-based U.S. monetary system which, as we have seen in our discussion of gold, lasted in various forms until 1971, when President Nixon ended international gold convertibility of U.S. dollars.

By the 1960s, the U.S. Treasury no longer issued "silver certificates," paper money secured by silver, and in all U.S. coinage minted after 1970, the silver content was eliminated entirely. Most other nations, with some notable exceptions like Mexico, have also ended official silver coinage production for monetary circulation, although a number do produce silver bullion coins for investment buyers.

THE PRICE OF SILVER

Although silver and gold were both important monetary metals, silver has now has lost this once historically dominant role.

Less a financial asset now and more an industrial commodity, silver's former use is more than equalled by its many newer fabrication uses, which continue to expand with advances in technology requiring electrical conductivity and noncorrosive metal alloys.

Nevertheless, the forces that influence the price of gold also have an impact on silver prices—diminished confidence in paper currencies, concern over prospective inflation, and political and economic turmoil around the world. It is no accident that silver prices, as a measure of precious metals, tend to follow and mimic the direction of gold prices, as does platinum.

There are four fundamental factors to watch in the silver market: total annual new supply, fabrication demand, investment demand, and existing inventories.

In recent years, trends in total new supply and industrial demand have been favorable to silver prices. Since the mid-1980s, annual new supply has decreased while photography, electrical, electronic, and other fabricated uses rose on average nearly 5 percent. In the emerging industrial nations of Asia, with 60 percent of the world population, demand has been much higher, approaching 17 percent in some years. Demand has been particularly strong in Taiwan, Hong Kong, South Korea, and Thailand.

On the other hand, net investor demand has lagged during this period, with many selling their existing stocks, and total old inventories have remained relatively large. Both factors have led to weaker silver prices recently, but this drop to some degree has been neutralized by low new supply and increased fabrication demand.

There has been a continuing imbalance in world silver production, in which new production in 25 of the 44 years between 1950 and 1994 has failed to meet actual demand. For example, in 1993, for the fourth consecutive year, fabrication demand exceeded total new supply by 165.9 million ounces. In 1992 the deficit was 63.2 million ounces. Existing stockpiles were drawn down sharply as a result. The cost of recovery of silver from scrap is so low that the flow of metal from this source continues no matter how low the price goes. Total 1994 demand was projected to exceed supply by 193.8 million ounces.

Some of this decreased silver mining production mirrored

SILVER

downturns in the prices of copper, lead, and zinc, of which silver is a by-product. This caused mine closings in Mexico, the United States, Peru, and Chili. Generally during recessionary periods, decreased industrial demand will drive down silver prices, one of the unfortunate aspects of silver's dependence on fabrication as its main use. When the economy lags, base metal prices drop, mining is curtailed, and silver as a byproduct stops.

Conversely, when silver prices have increased to highs of $6 or more an ounce, industrial demand has lowered historically as many fabricators stayed out of the market. Along with gold and platinum, silver's all-time modern high was reached briefly in 1979 and 1980 when it soared to $50 an ounce, but the price declined more than 90 percent from this peak throughout the 1980s.

Fabrication demand for silver rose almost 15 percent in 1993, much of it resulting from strong growth in silver jewelry and decorative objects in India and the Far East and from renewed silver coinage by Mexico.

The price of silver averaged $4.30 per ounce in 1993, up 9.4 percent from $3.93 in 1992. In late 1994 it was trading at $5.20 an ounce. In real 1950 dollar terms, silver prices averaged 73 cents in 1993, up from 68 cents in 1992, the first increase in annual average inflation-adjusted silver prices since 1987.

While some primary silver mines are still profitable, since 1990 many have closed, waiting for higher prices before reopening.

The real price of silver, when adjusted for changes in the Consumer Price Index since 1967, presents a sad picture. In real purchasing power, silver is nearly 20 percent below its 1967 price of $1.29. Adjusted for inflation, even with recent price rises, silver is at a 45-year low. As they say (and they are correct), silver has no place to go but up.

A Volatile Market

Even though silver is not on a par with gold as a safe haven against currency inflation, and taking into account its price swings, silver still offers some spectacular profits for those familiar with trading techniques and market trends. This is not an area for the inexperienced, but for those in the know, profits of 70 percent or more resulting from short-term price changes are not unheard of, particularly during periods of market weakness. A lot of profit was taken in 1979 and 1980 when the price of silver went to $50 an ounce.

The silver futures markets in New York, Chicago, and Tokyo offer the best potential rewards, as well as commensurate risks. Any commodities broker can arrange the purchase of a contract for the delivery of 5,000 ounces through COMEX in New York, or for 1,000 through the Chicago Board of Trade. Trading volume on the New York COMEX rose more than 18 percent in 1993, the highest level of silver futures activity since 1984.

According to the experts, much of this activity was caused by silver "hedge funds" and futures speculators. As with gold, buying silver "futures contracts" can protect large industrial users ("hedgers") from adverse fluctuations in the price of silver by guaranteeing them a steady supply at an established price. These hedgers—companies involved in the production, processing, or marketing of silver, including miners, smelters, and fabricators—employ silver future contracts as a practical tool for financial and planning management. In 1994 there was heavy investment in silver by these hedge funds, even rumors of hoarding of physical supplies in anticipation of a price rise to $6, $7, or more an ounce, at which point they would sell and take profits.

As with gold, direct purchases of silver may be made through banks, brokers, and precious metals dealers. Many people buy bulk lots of so-called "junk" coins—U.S. issues prior to 1963 when regular legal tender silver issues ceased. Also like gold, silver is sold in bullion bars with officially listed brands or markings. A 1,000-ounce silver brick is about the size of a loaf of bread and weighs about 68.6 pounds. Smaller quantities of 100 ounces, 10 ounces, and 1 ounce are also available.

A third, more indirect method of silver investment is buying silver mining stock shares, but their price tends to be rather static most of the time until general silver demand increases pro-

SILVER

duction. One of the oldest and largest U.S. companies is Coeur d'Alene, well-run and efficient, but its shares tend to be priced about the same as silver per ounce.

SILVER COINS

As with gold, there is a thriving market for rare silver coins, and the same factors tend to influence price and value: the actual precious metal content; the rarity, scarcity, or perceived collector's value; demand; and esthetic appeal and the state of preservation of the individual coin. Bullion coins are limited-mintage coins created by governments for the purpose of precious metals investment.

United States

In 1986 the U.S. Mint produced the first regular-issue gold and silver coins in many years. We have already discussed the gold American Eagle in the section concerning gold bullion coins, but there is also a 1-ounce silver half-dollar known as the "Walking Liberty" design. This original design was produced at the mints in Philadelphia, Denver, and San Francisco from 1916 to 1947 for general circulation and reintroduced for the 1986 silver bullion series of coin issues. The last regular-circulation silver U.S. coin was the Benjamin Franklin half dollar minted at Philadelphia from 1948 to 1963, but 40-percent silver half dollars were minted from 1965 to 1970. The 90-percent silver Kennedy half-dollar was minted in 1964 only.

All the silver used in American coin production comes from the nation's "National Defense Stockpile" and not from the open market.

Special note should be made of the series of U.S. limited-issue commemorative silver half-dollar coins, which were first minted in 1892. Since then a wide variety of historic events and American personages have been honored, usually with issues of 30,000 coins or less. Similarly, 13 commemorative gold coins were issued by the Mint from 1903 to 1926. It is estimated that less than 2,000 collectors have a complete set of these valuable U.S. commemorative coins.

Australia

The only major, legal-tender, pure silver coin to change its annual design is the 1-troy-ounce Australian Kookaburra dollar. It is 99.9-percent purity and weighs 31.365 grams. Only 300,000 are minted annually.

Mexico

In 1993 the Casa de Moneda, the national mint, continued to produce the silver bullion 1-ounce Libertad coin but also introduced several general-circulation silver coins, including those for 10 and 20 pesos, as well as a new 50-peso coin. As a mark of their popular value, Mexican officials estimated that 50 percent of the 20-peso silver coins and 80 percent of the 10-peso coins are being hoarded rather than circulated.

Canada

The Royal Canadian Mint's "Silver Maple Leaf" 1-ounce bullion coin has always been a popular seller, and sales topped $1 million in 1993.

INVESTORS AND THE INVENTORIES QUESTION

One of the former problems connected with silver investment was the unavailability of reliable statistics regarding supply and demand. This has now been largely corrected by publications such as the *World Silver Survey*, produced annually by the Silver Institute of Washington, D.C., a worldwide association of miners, fabricators, and manufacturers of silver. The Institute's address is 1112 Sixteenth Street N.W., Suite 240, Washington, D.C.

Nevertheless, there is an open question as to what the total existing inventories of available silver may be from all sources, new and old. Bearish comments are heard, without any particular

SILVER

substantiation, that there are at least 1 billion ounces of above-ground silver stocks available to meet demand. If nothing else, this assertion has had the impact of keeping silver prices lower than they might otherwise be.

The Silver Institute points out that the 1-billion-ounce figure was derived from estimates of persistent surpluses between 1979 and 1990.

When the price per ounce rose from $5 to $50 in 1979–80, it was assumed that all available supplies would have been sold at the increased price, and that unsold remaining stocks were not particularly price-sensitive. Similarly, when the price per ounce slid 90 percent during the next decade, no major dumping occurred, another indicator of illiquid silver holdings.

The actual 1979–1990 supply/demand balance totaled about 1.1 billion ounces, which was drawn down to about 810 million by the end of 1993, a reduction of about 8 percent a year. At this rate the predicted 1994 year-end available balance could be about 620 million ounces, a 25-percent decline from 1993. Meanwhile, on-hand manufacturing silver is at the lowest level since 1973, about 30 percent of projected annual use. Bank, investor, and dealer stocks have also declined and are now equal to just 42 percent of total use, the lowest level since 1982.

All of these factors, including available supply, seem to point to a slow, long-term increase in silver prices.

Most investment experts see silver as a conservative, long-term buy—one with significant speculative opportunities during the 1990s.

PLATINUM

Of the world's three leading precious metals—gold, silver, and platinum—in many respects, platinum is the least understood and appreciated. Gold and silver, widely used as monetary and investment metals for thousands of years, are firmly fixed in mankind's collective psyche as historically valuable commodities.

Say the word "platinum," and Americans are likely to conjure up a mental picture of some sort of special credit card, the trademark hair color of blonde movie stars, or maybe a rock star's million-selling recording.

This widespread lack of awareness is curious, because knowledgeable investors know platinum as a metal even more valuable than gold itself—a material with a bright future, both in multiple uses and return on investment. More than 20 percent of all manufactured goods now contain platinum or use it as an essential part of their production process. In 1980 when the price of gold hit an all-time high of more than $800 a troy ounce, platinum topped out at more than $1,000 per troy ounce.

However limited the U.S. general public's perception may be, no serious modern investor can afford to be unenlightened about platinum's potential.

THE EARLY HISTORY

Platinum is an attractive silver-white metallic chemical element, dense, inert, ductile, highly resistant to corrosion, with an extremely high melting point (1,774 degrees Celsius)—higher even than gold. It is one of the six transition elements in Group VIII of the periodic table, known collectively as platinum group metals (PGM), including ruthenium, rhodium, palladium, osmium, and iridium.

Platinum is extremely rare, with an abundance in the earth's crust of only about one millionth of one percent. All the platinum ever mined would easily fit into a square space of less than 25 feet to each side—about the size of a shipping container for a BMW.

Platinum's symbol is **Pt**, its name derived from the Spanish *platina*, literally meaning "little silver," due to the obvious resemblance noticed by the conquistadors who first encountered it. The French-Italian physician Julius Caesar Scaliger in 1557 recorded the first known allusion to a refractory metal, probably platinum, found in the region between Darien (now Panama) and Mexico by Spanish explorers seeking gold.

The metal was used by pre-Columbian South American Indians well before it was identified in 1735 by the Spanish mathematician Antonio de Ulloa in alluvial deposits combined with gold at the mouth of what is now the Rio Pinto in Colombia. Independently in 1741, English metallurgist Sir Charles Wood distinguished platinum in ore samples he obtained in Jamaica. By 1751 a Swedish scientist, H.T. Scheffer, declared it to be "the seventh metal known" to man, the others then being gold, silver, copper, tin, lead, and iron.

Valuable from the Beginning

Platinum's earliest "practical" use was as an adulterant of gold by crafty Spanish counterfeiters who learned that, when combined, the resultant alloy was virtually indistinguishable from pure gold in color and weight. A threat to the gold escudo,

Platinum

the practice was so widespread that the Spanish monarch imposed a penalty of death on those found with platinum in their possession. When that Draconian measure failed, the King offered an exchange of two ounces of silver for each one of platinum—the first known official evaluation of the metal. Later, as fiscal problems beset the Spanish throne, the royal mint secretly was ordered to use platinum for the same illegal debasement purpose, and this diluted "gold" was used to settle pressing foreign debts, a possible precursor of today's international foreign exchange system.

Spain made this mintage policy official with the first minted platinum currency, an eight-escudo coin issued in 1747. France struck some in 1801, and the London Royal Mint issued a platinum penny in 1807.

In 1802, English chemists discovered "platina" was in fact not a single metal but a natural alloy of several metals that came to be known as the platinum group metals, the most plentiful of which they named "platinum."

Presaging modern usage, 18th-century European metal workers employed platinum in the manufacture of such diverse objects as jewelry, religious artifacts, cutlery, and heat-resistant crucibles and evaporating dishes needed for chemistry. One of the more elegant craftsmen to employ platinum was the royal goldsmith to Louis XVI of France, one Marc Etienne Janety, a rare surviving example of whose work—an ornate sugar bowl lined with deep blue glass—is now exhibited in the Metropolitan Museum of Art in New York.

Ruble Troubles

With the discovery of rich placer deposits in the Ural Mountains in 1824, Tsarist Russia took the lead in platinum production, as South American deposits dwindled.

Aware of the unfortunate Spanish gold escudo experience, the Imperial Mint in St. Petersburg made every effort to control all platinum supplies, even producing its own jewelry—which wealthy Russians rejected out of hand as "imitation silver."

PORTABLE WEALTH

Platinum coins in three-, six-, and twelve-ruble denominations were minted as legal tender beginning in 1828, with respective weights of 10.55, 20.7, and 41.4 grams. The Russians used more than half a million ounces of native platinum making these coins before withdrawing them in 1846, when the price of platinum dropped drastically. The discovery was made that for 20 years, European counterfeiters had grown rich using "worthless" Colombian platinum to duplicate the Tsar's coins, more than doubling the total official circulation.

MODERN USES OF PLATINUM

By the 1890s, science had learned much about what would become platinum's premier industrial role—its extraordinary catalytic properties—that is, its unequalled ability to alter and/or accelerate numerous beneficial chemical reactions.

In the 1920s, the International Nickel Company first produced platinum as a by-product of its refining, and subsequently major sulphide copper-nickel ore deposits containing PGM were found in the Talmyr Peninsula of Russia and in the Sudbury district of Ontario, Canada. It was also in the early part of this century that a consulting geologist in South Africa, Hans Merensky, correctly predicted the existence of a massive subterranean reef of platinum in an area known as the Bushveld Igneous Complex in the Transvaal. By the 1960s, the Merensky Reef mines, the first ever worked primarily for platinum rather than other ores, became the world's principal source.

Expansion since has made South Africa dominant in worldwide production, although Russia, which still does not release its production statistics is also a source of major nickel by-product, much of it used internally until 1990 when platinum exports rose sharply as a means of earning hard currency. More recent declines in world nickel demand have consequently reduced Russian platinum production.

Today, in its finely divided form, about half the total platinum production serves catalytic purposes, much of it in auto-

Platinum

catalysts and in the petrochemical industry for gasoline cracking (extraction from petroleum) and the manufacture of high-octane fuels. Platinum catalysis in the chemical industry also produces such substances as the widely used nitric acid, essential in the production of fertilizers, explosives, dyes, some plastics, and medicines.

Fine platinum wires glow when placed in alcohol vapor, converting that substance to formaldehyde, the same process that allows its use in cigarette lighters and heaters. With a coefficient of thermal expansion similar to that of soda-lime-silica glass, platinum is used to make sealed electrodes in soft glass systems. So-called thermocouple sheaths use these thin platinum wires for temperature measurement in steel, glass, and semiconductor manufacture. It is also used for producing fiberglass and high-quality optical glass and fiber optics.

Also drawing on platinum's superb catalytic properties, Universal Oil Products Company developed the so-called "platforming process," opening the way to huge growth in the production of synthetic fibers, plastics, insecticides, styrene, polystyrene, nylon, polyester fibers, and many more petroleum-based products. Platinum also plays another important role in textiles, where the microscopic holes in platinum nozzles make practical the spinning of synthetic fibers, and similarly in jet-engine fuel injection nozzle systems.

Although it does not consume large quantities, the medical use of platinum is essential because of its compatibility with all skin types and allergy-free properties. It is used for heart pacemakers and catheters, electrodes used for monitoring patient conditions, mass spectrometers, dental technology, and neurosurgical equipment such as pins used in hip replacement and other bone surgery. Second-generation platinum component anticancer drugs include Paraplatin, used in treating testicular, ovarian, and head and neck cancers. The widely used inhalant Ventolin is also a platinum-based medication.

A newer, growing field of platinum use, only now in its infancy, is in "fuel cells," self-contained energy units that produce electricity, heat, and water by chemically combining hydrogen and oxygen. Mazda already has in operation a prototype platinum-based "proton exchange membrane" fuel cell able to power an auto. Platinum is the indispensable catalyst. Already standard equipment in American space shuttles, fuel cells are expected to play an expanding role in nonpolluting domestic and commercial power production and in motor vehicles approaching an eventual "zero emission" standard. Experimental fuel cell-powered buses and autos are already on the road. The future demand for these products could be enormous—and they all require platinum.

Similar industrial emission control of toxic gases is being accomplished by large-scale platinum-based catalysts installed in factories and utilized in businesses ranging from paint, coatings, food processing, plastics, and utility plants to dry cleaning establishments and bakeries. Under the Clean Air Act of 1990, all of these U.S. businesses have been subjected to even stricter emission controls, none of which can be met without platinum.

Clean Air for the World

Perhaps platinum's best known contemporary use (40 percent of 1994 total annual production) is for the "auto catalyst"—the antipollution catalytic converter that is now a part of automobile exhaust systems in nations all over the world. First used to meet California state laws mandating emissions standards in the 1970s, by 1973 these devices were used to meet federally mandated emissions standards. The standards were further strengthened by the Clean Air Act of 1990. Other nations have followed suit, including Japan, Singapore, Hong Kong, Mexico, and the countries of the European Union (EU). At least half of the world's autos and trucks are now required to be outfitted with platinum converters.

Instead of spewing deadly carbon monoxide, hydrocarbons, and nitric oxide into the atmosphere, platinum converters oxidize or reduce internal combustion auto engine exhausts into relatively harmless substances like nitrogen, carbon dioxide, and

PLATINUM

water, assuring much cleaner air now and in the future. Demand for platinum used for this purpose has now reached 1.5 million ounces per year and is growing because of increasingly tighter emission standards taking effect in the United States and other nations each year. These original U.S. standards were adopted in the European Union nations on January 1, 1993. European standards are tightening further and will come up to 1994 American levels by 1996. As a reflection, European sales of platinum autocatalysts surged 65 percent from 1990 to 1994, as the new standards were phased in.

The use of platinum in auto catalysts is estimated to have risen by 160,000 ounces in 1994 in response to growing U.S. auto production, pent-up demand for new cars, and tougher emissions standards. With the simultaneous expansion of the world economy in 1994, the worldwide auto industry still bought about 1.84 million ounces, surpassing a previous annual high of 1.68 million set in 1992. The continued growth of the world economy, including greatly increased auto sales, can only increase platinum and PGM (palladium and rhodium) demand further.

In America new car sales have been rising for three years, one of the major factors being the average age of cars already on the road—8 years, the oldest average since 1948. Simple demand for new cars means more platinum consumption, as do more stringent environmental laws.

THE ESTHETICS OF PLATINUM

Complaints about "imitation silver" evidently were forgotten by the turn of this century. Platinum had achieved acceptance and popularity as a desirable jewelry metal. In 1905, the price of platinum per troy ounce topped $20, the first time it exceeded that of gold, much of the demand generated by jewelry manufacturers.

Platinum jewelry gained great favor in Europe and the United States as the unrivaled material of the 1920s Art Deco period. It was the fashionable Parisian House of Cartier that created the first wristwatch using platinum.

Though the 20th-century demand continues, it is hardly a new phenomenon. Platinum and gold jewelry have been found in pre-Christian Egyptian tombs and in pre-Colombian areas of that Latin American nation and Ecuador.

Today, fully 40 percent of platinum production goes into jewelry, much of it sold in Japan, where a special affinity for the metal has developed, especially among affluent younger age groups. Japanese jewelers have produced a wide variety of beautiful platinum and gold pieces, prized not only for their purity of style, but also their intrinsic worth; a 30-year low in the yen price of platinum certainly influenced this buying trend. In 1994 European jewelry demand decreased modestly by about 10,000 ounces, as U.S. demand increased by about 15,000 ounces.

Unlike gold, platinum jewelry and coins require no karat alloying to enhance hardness and durability. As compared to gold jewelry, which often contains as little as 50-percent gold, platinum jewelry usually is 90- to 95-percent pure. Platinum coins and bars now have a .9995 purity.

Platinum is a metal with unique properties suitable to personal use as jewelry. It is compatible with all skin types, nonallergenic, does not discolor or tarnish, is heat resistant, wears very well, and is an ideal setting for precious stones, which it enhances because of its more neutral color.

Over 1.7 million ounces of platinum were used in jewelry manufacture in 1994, an increase of more than 85,000 ounces over total 1993 use—roughly 85 percent of it going to Japan.

PLATINUM AS AN INVESTMENT

Platinum is traded in all the world's major bullion markets in response to popular demand for the metal in its pure form. Legal tender coins and small bars continue to consume a significant percentage of total production each year. Depending on the current market conditions, annual use for investment purposes can average from 5 to 15 percent.

The North American platinum physical trade is based in

Platinum

New York City, and futures trading occurs at the New York Mercantile Exchange (NYMEX), the first platinum futures market in the world. Principal European markets are in Zurich and London, where bars are traded in lots worth about $350,000 each. As with gold, there is a morning and afternoon London price fixing of platinum, conducted by the London bullion houses and three major Swiss banks, to balance immediately available supply and demand. In the Far East, trading activity occurs in Hong Kong and Tokyo, mostly in small bars and coins, which are extremely popular in Japan. Tokyo has a futures contract on platinum.

Platinum Compared to Gold and Silver

There have been wide swings in the price of platinum over the last 15 years, not unlike those of the other precious metals, gold and silver, which move in tandem with platinum. However, the relative smallness of the platinum market makes its price more volatile, thus bringing greater returns during upward price movements. Compared to total platinum output, more than 10 times more gold and 125 times more silver is produced each year.

In 1960, with the U.S. government still pegging gold at $35 an ounce and forbidding private ownership, much was made of the fact that platinum, for the first time, traded at more than $100 an ounce. However, capturing a share of the precious metals market proved very difficult in competition with more traditional gold and silver investments.

During the gold-silver bubble in 1980–81, platinum traded at more than $1,000 an ounce when gold was over $800, a sevenfold price increase in just five years. After that peak, the price fell for five years to a low of $237, then doubled again in three years to $680 in 1986. In the late 1980s platinum prices assumed a more realistic comparison to available supply and demand factors. In 1993, its high was $419 in August, the highest price in two years, and the low was $339.75 earlier in March. The yearly average was $374.06, up 4 percent from the 1992 average. During 1993, platinum led gold by a spread of about $40 an ounce. By comparison, in the last half of the 1980s, platinum led gold by $100 an ounce.

The available supply of platinum is far more finite than gold, and thus it has a unique intrinsic value. Although platinum has a limited role as legal tender coinage, many nations, including the United States, hold it in their strategic stockpiles because of its importance in industrial and military uses. Platinum and gold differ in that platinum's value depends on industrial uses far more than gold. Gold is primarily a monetary metal and a financial asset.

Both the fundamental differences and the similarities between the two metals make platinum an excellent complement to gold. Indeed, if an investor has cultivated a winning gold investment strategy, the same approach will likely be successful with platinum, but with a good chance of even greater returns because of the greater volatility of the platinum prices. The historical fact is, platinum tends to outpace gold in bull markets.

Silver, as a widely used industrial metal, has some important similarities to platinum but, as we have seen, there is a major difference—there are relatively large supplies of silver available above ground, while platinum stocks are limited in source and availability. Because they are both primarily industrial metals, the price of both usually mirrors the directional trends of the general world economy. Fluctuations in silver therefore can be instructive about the price of platinum, discounting the difference in stocks.

For a long time the conventional wisdom was that platinum followed the price of gold since both precious metals were said to be subject to the same operative price factors. Most experts now admit that platinum has a set of unique influential price fundamentals all its own, built on growing industrial demand. Therefore, platinum is capable of setting its own independent price level in the free market, usually at a decided premium to gold.

Because of the high cost and lengthy time period of platinum production, total output cannot readily increase to meet rising demand. Increasing supply requires considerable fore-

PLATINUM

sight and planning because the tap cannot be turned on overnight, due to technically difficult mining and refining operations. This means that when demand rises, prices will also rise, making platinum an investment with a limited downside risk and considerable upside potential.

One point is very clear: just as gold and silver are real commodities that wise investors purchase as a hedge to diversify their paper assets, platinum provides even greater diversification within that diversification, but with a rewarding kicker—a consistently higher market value as reflected in the price per ounce.

Investment Opportunities

As with gold and silver, methods of investing in platinum are readily available in various forms—futures and options contracts, coins, bars, and certificates. So far the most popular investment avenue has been purchase of platinum in its pure form.

The most immediate and convenient way to invest is to purchase legal tender bullion coins. In 1980, only 250,000 ounces of platinum were held worldwide in small investments forms, but today more than 3 million ounces worth over $1 billion are held by investors. Small investment forms include legal tender coins, medallions, and bars weighing 10 troy ounces or less. Large investment forms include 500-gram and 1-kilogram bars (about 31 troy ounces), which are popular in Japan, and platinum held on account for subscribers to accumulation plans.

Although many nations have issued platinum coins since the first Spanish escudos were struck as early as 1747, more recent modern issues began in earnest in 1975:

- **Isle of Man:** This semi-independent island community within the United Kingdom began issuing platinum year sets in 1975 and in 1983 issued the "Noble," which is sold internationally. These coins are .9995 pure, as are those issued by Australia and Canada, the three national mintages that are the most commonly traded.

- **Australia:** In 1987 Australia began issuing annual platinum uncirculated bullion coins in limited amounts. The 1989 platinum Koala was struck by the Perth Mint in one-twentieth-, one-tenth-, one-quarter-, and one-half-ounce sizes sold in sets, but the one-twentieth-ounce coin can be purchased individually.
- **Canada:** In 1989, the Royal Canadian Mint marked the 10th anniversary of the Maple Leaf bullion coin program by issuing a 1-ounce platinum design. In 1991, platinum coins were struck in $30, $75, $150, and $300 denominations featuring snowy owl designs.

Other nations have followed suit in issuing various legal tender platinum bullion coins, including Singapore, Portugal, the Russian Republic, China, Liberia, Tonga, Bhutan, and Sierra Leone. In addition, there have been numerous mintages of official and unofficial commemorative bullion coins by France and Italy and by private companies in Japan, Hong Kong, and other nations.

Worldwide demand in 1993 for small investor forms of pure platinum was about 125,000 troy ounces, valued at $47 million. In 1994 the demand rose to about 150,000 ounces as investor interest grew, especially in the United States, where inflationary fears led investors to the security of tangible assets.

The advantages of investing in small platinum coins are many, including an existing international market, gradual build-up of a portfolio, easy portability, and ease of identification and authentication. Commissions on platinum coin purchases are usually not charged, with dealers making their profit on the dealing spread.

Some firms have begun offering platinum accumulation plans that are similar to their gold counterparts. The Mocatta and Perth Mint storage certificate programs discussed in detail in the chapter on gold are also available for platinum investors. Full

PLATINUM

information on this program is available from Asset Strategies International, Rockville, MD.

Platinum coins have largely edged out platinum bars and ingots for small investors because of their comparative ease of purchase, sale, and storage. Investment-grade bars are available in 10-ounce and smaller sizes from various refiners.

In coins, metals, and bars, platinum is gaining popularity as a viable numismatic and bullion alternative and/or supplement to gold and silver.

There is also an active platinum futures market largely dominated by industrial hedgers, not usually a place for small buyers because of volatility and large participants.

Investment in platinum mining shares is also available, with several South African firms now trading over the counter in the United States and western Europe. Among these companies are Rustenberg Platinum Holdings and Impala Platinum. In late 1994 it was calculated that every 10-percent increase in the platinum price increased Rustenberg's earnings 45 percent and Impala's by as much as 60 percent. With all this activity in platinum, the third precious metal has come a long way from being a discard thought to be worthless.

PLATINUM AND CURRENCY FLUCTUATIONS

A word about a related topic: international currency markets. Changes in these markets can sometimes influence platinum's price because of fluctuations in the value of the U.S. dollar, the currency in which platinum trades worldwide. If the dollar price remains the same, then a weakening currency against the dollar means a rising value of platinum in that currency. When the dollar weakens compared to a stronger currency, investing in platinum becomes a very attractive alternative to the holder of the stronger currency. For example, a Japanese investor who paid 250 yen per dollar for U.S. Treasury bonds in 1982 received less than 100 yen per dollar when he sold those bonds in 1994. The same forces that produced this imbalance in the two currencies also makes reduced platinum prices expressed in appreciated yen highly attractive to the Japanese as a real commodity in which to invest—certainly far more attractive than U.S. government paper! This, in part, explains why 46 percent of the total 1994 world demand (4.635 million ounces) for platinum came from Japan, much of it for jewelry, and why, in spite of a recession during 1992–93, Japanese demand has increased in most years since 1983.

Fundamentals of Supply and Demand

As noted, since the mid-1980s, platinum production and consumption—supply and demand—have been the most influential factors in the international market, avoiding the speculative gyrations of the early part of the decade. While comparable price changes have not been so dramatic, the longer-term trend has been a consistent increase in platinum value.

Platinum has the most compelling long-term supply/demand fundamentals because 1) it has many essential and often irreplaceable applications, 2) over 95 percent of the world's platinum is mined in Russia and South Africa, and 3) above-ground supplies are limited.

It should also be noted, as an added factor, that the mining and refining of platinum is an enormous task requiring advanced technology, secure financial backing, and relative labor tranquility. Availability of the finished product requires the extraction of *ten tons* of ore from deep shaft mines just to produce *one ounce* of pure platinum. The entire process requires upwards of six months, and bringing a new mine into operation takes a minimum of five years, and usually closer to ten years.

Only about 140 tons of platinum reach Western world markets each year.

Working under these practical supply limitations, increased demand can only be satisfied relatively slowly, assuring continued strong prices in the long run for platinum and for investors.

PLATINUM

The Russian Situation

Although the Russian Republic produces 15 percent of world platinum, there are limiting factors that make this source potentially unreliable in the extreme, not the least of which is the general economic chaos in the former Soviet Union.

Not only has central government control slipped in many areas since the ouster of the Soviet system, but distribution, transport, and financial systems are acknowledged to be in varying states of collapse, perhaps none more so than in the area of mines and mining, a large and essential part of the Russian national economy. The result has been widespread labor unrest in Siberian and other mines, with strikes and the threat of strikes being a constant occurrence. Antiquated plants and equipment, lack of transport, surplus workers no longer subsidized by the state, layoffs, and inadequate capital all combine to make Russian nickel/platinum production chancy at best for the near future. Production, refinery, and distribution bottlenecks have already caused delayed and missed deliveries, especially at the Norilisk Nickel Combine (NNC) in Siberia.

Because Russian platinum is a by-product of nickel production, as noted, the fall in world nickel prices has partially led Russia to lower production of platinum since 1992. The existing limitations on the Russian economy make it virtually impossible for any wholesale conversion to pure platinum production, with the attendant cost of new equipment and technology. In spite of increased Russian platinum sales in the international market in order to gain much-needed hard currency, the continued availability of Russian platinum is certainly not guaranteed.

Platinum supplies from Russia rose in 1993 to 775,000 ounces, up from 750,000 in 1992.

The South African Question Mark

Which leads us to the highly fluid situation in the Republic of South Africa—the one nation that produces from 75 to 85 percent of total annual world production of platinum. It can be reasonably argued that if platinum supply disruption is a real potential in Russia, it is equally a possibility in South Africa. In the run up to the April 1994 elections that brought to power Nelson Mandela's government of national unity, world spot prices of platinum exceeded $430 an ounce, largely because of the concern produced by constant violence between Mandela's African National Congress and rival black factions, especially in Natal Province.

While observers have been pleased at the relatively smooth political transition to majority rule, the enormous economic expectations of blacks, especially the large contingent of Congress of South African Trade Union (COSATU) mine workers, hardly have begun to be addressed by the new Mandela coalition government. Instead of a fast start on new housing, social welfare programs and policies that impact the average person, the nation has been treated to a spectacle of black and other minority members of parliament and the government raising their own salaries, reveling in official perks and generally seeming to ignore their constituents' needs.

Strikes have already broken out in many sectors, including mining, with blacks protesting the existing built-in disparity between the wages of blacks and whites, as well as white dominance of management positions. While the new government's surprising pro-business and investment policy has pleased potential foreign investors, ANC diehards continue to call for massive new taxes on business to finance an accelerated social welfare agenda. Mandela's advanced age and the person who might succeed him are also wild cards that must be considered.

The answer to whether these internal issues can be addressed quickly enough to avoid political and economic chaos is unknown, and in that very uncertainty lies the potential for disrupted or decreased platinum supplies, meaning higher potential prices.

As a result of these imponderables in 1994, South African platinum supply fell from 1993 levels by 4 percent, or 120,000 ounces, to a total production of 3.24 million ounces. Nevertheless, all the mining companies have been forced to reevaluate future expansion plans as a result of declining revenues and increasing costs. Recently, several mining operations were closed and planned expansion at others delayed.

What any major disruption of South African platinum production could mean has already been documented. In 1985–86, a five-month strike at one of the "big three" mining companies, Impala, spurred on an already accelerating world platinum price from $235 an ounce to $682 in a 12-month period. That means a loss of only 12 percent of annual world production was a major contributor to a price increase of 190 percent!

The questions shrouding the future of the world's two major platinum supplying nations, especially in the face of an international economic revival with greatly increased auto sales in the United States and Europe, can only mean one thing: accelerating demand. There will probably be a small supply deficit and resultant higher platinum prices for the rest of the 1990s.

What may be bad news for platinum users could be very good news for those with sufficient discernment to make an opportune investment choice of platinum.

TAX STRATEGIES FOR PRECIOUS METALS INVESTORS

You may have purchased bullion coin investments in the 1970s and 1980s at prices substantially higher than today's levels. For instance, gold is now trading in the $400-per-ounce range, less than half of its historic high of more than $800 per ounce. Silver is also trading much lower than its previous peaks. These low prices in tangible assets may not last for long.

Because stock "wash sale" rules do not apply, Asset Strategies International, Inc. (Rockville, MD) has developed a program that enables you to take advantage of today's low prices to reduce your tax burden by taking losses on your coin investments to offset current ordinary income, or to shelter gains that you have realized on your other investments. Investors owning bullion coins that have declined in value can sell those coins to Asset Strategies, thereby creating a deductible loss, and will have the option, but not the obligation, to buy the same coins back from Asset Strategies or buy other, materially different coins or metals from Asset Strategies.

Provided that the transaction meets the criteria set forth in the laws and regulations, any loss resulting from your sale or exchange can be deducted up to $3,000 of ordinary income, or it can be deducted against capital gains on other investments, with

PORTABLE WEALTH

the ability to carry unused losses forward to future years. This deduction will be available even if you exercise your option to repurchase the coins from Asset Strategies.

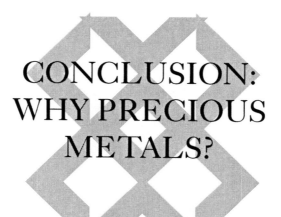

CONCLUSION: WHY PRECIOUS METALS?

History teaches us some excellent lessons, if we are just willing to learn. Unless the world becomes a dramatically better place in which to live, sensible people will continue to invest in real commodities with enduring value—precious metals foremost among them. Whether as a hedge against inflation; because of instability in currency markets, the mixed state of the stock market, and political instability; or prompted by a shrewd reading of supply and demand factors, smart investors are going to buy gold, silver, and platinum.

The dynamics of current supply/demand fundamentals, improving world economies, persistent political turmoil, a positive commodity cycle—all these are strong indicators of a possible bull market for precious metals during the 1990s, with platinum leading the way.

ABOUT THE AUTHOR

Adam Starchild is the author of more than a dozen books and hundreds of magazine articles, primarily on business and finance. His articles have a appeared in a wide range of publications around the world—including *Business Credit, Euromoney, Finance, The Financial Planner, International Living, Offshore Financial Review, Reason, Tax Planning International, Trusts & Estates*, and many more. His personal Web site is on the Internet at http://www.cyberhaven.com/starchild/